Jorge Alfredo González Pérez
Cesaltina Carla Da S. Mussinda

Agro-productive performance of beans with organic fertilizers

Jorge Alfredo González Pérez
Cesaltina Carla Da S. Mussinda

Agro-productive performance of beans with organic fertilizers

Agro-productive performance of beans(Phaseolus vulgaris L.) with organic fertilizers in the Caxila 3 district

ScienciaScripts

Cover image: www.ingimage.com

This book is a translation from the original published under ISBN 978-620-0-00961-6.

Publisher:
Sciencia Scripts
is a trademark of
Dodo Books Indian Ocean Ltd. and OmniScriptum S.R.L publishing group

120 High Road, East Finchley, London, N2 9ED, United Kingdom
Str. Armeneasca 28/1, office 1, Chisinau MD-2012, Republic of Moldova, Europe
Managing Directors: Ieva Konstantinova, Victoria Ursu
info@omniscriptum.com

Printed at: see last page
ISBN: 978-620-8-39901-6

THANK YOU

✓ First of all I thank Almighty God for protecting me throughout this great academic journey, then I thank deeply my family, friends, colleagues and all those who directly or indirectly made this great journey come to an end.

✓ I will not forget to thank the collective of professors who did so much to enable me to achieve this great merit. These thanks extend from the first year to the last year, especially to the master's professor. Jorge Alfredo Gonzalez Perez, who is the great responsible for this work, thank you all very much, God bless you richly and powerfully.

To all my deepest thanks.

DEDICATION

√ I dedicate this work to the very important people in my life, who are my mother, Mrs. Deolinda Carlota Melgaço da Silva, who is responsible for my existence on the face of the earth;

√ The memory of my father;

√ To my children for being the reason I face every battle;

√ To my brothers who always accompanied me at every moment;

√ to them and for them I give this small gift, because I know they deserve much more.

SUMMARY

With the purpose of evaluating the effect of different doses of organic fertilizer on agro-productive indicators in bean crops (Phaseolus vulgarisL) as indicator plant in a mine in the locality of Caxilla III, municipality of Cuanhama, Province of Ondjiva, this experiment was carried out in the months between February and June 2023. The experiment was carried out in open field conditions where an experimental design of randomized blocks in a factorial scheme of four treatments each with three replicates was performed. The treatments used were organic fertiliser from bovine manure (5.0, 10.0 and 30.0 kg.m^2), and a testimonial without application. The variables evaluated were: plant height (cm), stem diameter (mm), number of leaves, number of pods per plant, pod length, weight of 100 cements and yield per plot. The data obtained were subjected to analysis of variance using the statistical programme Statgraphics 5.1. For the economic valuation, the following economic indices were calculated: production cost (Akz), production value (Akz) and profit. Secondly, the best treatment (T3) was obtained when organic fertiliser of bovine manure was applied at 30.0 kg. m^2 , which influenced plant height, stem diameter, pod length and hundred grain mass. Economically the treatment when organic cattle manure fertilizer was applied at 30.0 kg. m^2 (T3) was the best profit result of the experiment.

Key words: Fertilisation, treatment, ploughing, organic.

INDEX

I. INTRODUCTION

Agriculture and its productions represent the basis of sustenance for most of the world's populations, its scientific advances and technological processes depend on increasing the quality of life on the planet; meeting the food needs of a population that is growing at a dizzying pace and the essential challenges of the 21st century, in the future these needs will have to be guaranteed fundamentally by increasing crop production, especially those that constitute the food base of each country. region (Altieri, 1999 cited by Namwenyo, 2017 and cited by Pequenino, 2018).

Beans are a major source of plant-based protein with wide global demand, as reflected in data from FAO (2011), assessing the high levels of marketing and consumption. The food security of this important component of the diet has been a concern for the state, the panorama of the production systems of this crop has been experiencing a constant instability, impacted by several reasons, during the period from 2008 to 2010 the areas dedicated were: 76,740 a
112,201 ha, and yet production ranged between 70,600 and 132,900 tonnes, for a yield of 0.91 to 1.18 t.ha-1 , these results have been insufficient to guarantee the consumption of the population, being forced to make constant imports on the versatile market, and which, with its price fluctuations, constitutes a drain on the State's finances. Suárez; Mamani; Joaquín (2018) and cited by Pequenino (2018).

The bean (Phaseolus vulgaris L) belongs to the large legume family and its origin dates back approximately 7,000 years. According to the classification of centres of origin of cultivated plants, described by Binder (2007), this species is located in genocentres VII and VIII, which cover the Central and South American regions of Mexico (including the West Indies), South America (Peru, Ecuador and Bolivia).There are currently 180 species of the genus Phaseolus in the world, of which approximately 126 are from the Americas, 54 from South Asia and East Africa, 2 from Australia and only one from Europe (Box, 2011). Beans are a well-known crop grown throughout much of the world and are widely used as food and feed, due to their nutritional qualities. There are several species and varieties of this crop, all belonging to the genus Phaseolus. All scientific evidence suggests that they are plants of American origin, as they have been cultivated there since pre-colonial times (FAO, 2015).

Beans have high production potential and respond well to the use of technology. Its cultivation is generally mechanised, benefiting greatly from modern planting and harvesting techniques (Box (2011). Annual bean grain production is approximately 3.6 million tonnes, approximately 64% of which comes from West and Central Africa (Baudoin et al., 2021).

Bean crops are one of the staples of the Angolan diet. Not only as a source of vegetable protein, but also as a source of iron, fibre, folic acid, thiamine, potassium, magnesium and zinc. These compounds are of vital importance in the prevention and treatment of cardiovascular diseases, diabetes, obesity and colon cancer, among the most common.

The consumption of food with a high iron content reduces anaemia in children. In the case of Angola, annual production is 3,311,988 tonnes with an area of 934,947 ha and in the Cunene region it is 334 tonnes with an area of 4,712 ha (MINADER, 2015). In Angola, a project financed by the United Nations Development Programme (UNDP) and carried out jointly with the Angolan Agricultural Research Institute has been working since 2012 to improve the living conditions of six hundred farming families by growing beans.

To this end, according to FAO (2013), they were strengthened with five cooperatives located in the provinces of Huambo and Bié, where they created field schools that allowed them to monitor the planting of 64 hectares, improving the farmers' agricultural practices. It is a highly productive and hardy crop, which allows it to be cultivated at various times of the year.The studies carried out refer to the adaptability of beans to soil and climatic conditions Sandoval (2014) says that cultivars have intrinsic genetic, physiological and morphological characteristics and therefore respond differently to local soil and climatic conditions, highlighting the importance of studying the productive behaviour of cultivars before. recommending cultivation.

Genotype-environment interaction effects can be the result of different factors, such as: environmental conditions, fertility of the soil, the technological knowledge of the farmers and the management system adopted. These factors, alone or together, can change the performance of a genotype in a region (Ndanyengwondunge, 2014).Phenotypic expression in plants of different bean varieties depends on their genetic content and environmental interaction. Whereas, for the same variety, the greatest agricultural benefit may vary according to the growing conditions. It is necessary to adjust the programmes for each variety to the conditions present in the agricultural areas. Any value that has not been tested is not reliable. The study of the influence of soil and climate on the growth and phenological development of plants of different crop varieties is fundamental to evaluate these results. Over the years, bean cultivation has not reached an acceptable level of production in Angola, especially in the province of Cunene, where climatic and soil conditions play a determining role in the good physiological development of this crop.The food security of this important component of the diet of the Angolans and in particular of the Cunenesses, has been a concern of the State, the panorama of the production systems of this crop has been in constant instability. Knowing the importance of the crop for food and that in the province of Cunene it is cultivated by farmers, since the climatic conditions are not very favourable for its development, taking into account the average annual rainfall that varies between 600 mm in the months of November to April, with predominantly high temperatures of 30 to 35 ºC (Cunene Meteorological Station 2023). One of the production problems is that there is neither the necessary knowledge of the behaviour of the improved cultivars for their establishment, nor the amount of seed required for these varieties to meet the needs of local producers for the implementation of bean cultivation in the locality. Another problem that persists so far is the use or extension of good organic allocation practices applied to the different crops that need to be grown in this area of Angola. Taking into account the various issues raised above, it is defined as follows the following scientific problem for this research.

Problem: What will be the morphoagronomic behaviour of different sources of organic fertiliser applied to bean crops sown in a family garden in the Caxilla III neighbourhood?

On this basis the following hypothesis can be granted:

Hypothesis: If different sources of organic fertiliser were evaluated in the cultivation of beans (Phaseolus vulgaris L), then the effect on the agro-productive indicators of the plant in a family garden in the Caxilla III neighbourhood could be determined.

We set out the following general and specific objectives of our research as follows:

General objective: to evaluate the agro-productive effect of different sources of organic fertiliser applied to the cultivation of beans (Phaseolus vulgaris L) as an indicator plant sown in a family garden in the Caxila III neighbourhood.

Specific objectives

- To determine the response on morpho-physiological indicators triggered by fertilisers of the bean crop (Phaseolus vulgaris L.) as an indicator plant.
- Economic valuation of fertilisers applied to the crop.

bean (Phaseolus vulgaris L.) as indicator plant.

II.BIBLIOGRAPHIC REVIEW

2.1.Importance of legumes

Legumes are of great economic importance as they produce high yields and a large proportion of nutritional reserves, the application of which to human or domestic animal feed has occupied and continues to occupy a place in agricultural practice (Exania and Zoraida, 2006).

Although the primary utility of grain legumes lies in their seeds, these plants also have multiple uses in agriculture, for example as green manure, fodder and silage (Sandoval 2014). Green manure are all those legumes that are used to improve the soil; their use increases soil fertility, preserving it against erosion and conserving moisture during the dry period and controlling weeds.

Legume residues as a source of nitrogen for the following crop are much more important than cereal residues. Legume residues in general are good fertilisers because of their higher nitrogen content, this nitrogen is assimilated more quickly by other plants and has a C:N ratio of less than 30:1, therefore it tends to release nitrogen and decompose it quickly (Sandoval. , 2014).

The main advantages of using leguminous cover crops on crops are weed management, maintaining a moist soil layer, increasing organic matter content, nitrogen fixation and soil protection to prevent erosion (Arias et al. al., 2009).

2.1.1. Origin and importance of the beans

The cultivation of beans (Phaseolus vulgaris L.) spread to almost all pre-Hispanic populations and constituted one of the main food crops, along with maize, potato and cassava (Escoto, 2014).

It is a plant native to Central America and southern Mexico. Cultivated in ancient times, it can still be found spontaneously in South America. It was brought to Europe shortly after the discovery of America and subsequently, cultivation became increasingly important according to its adaptability, spreading throughout both hemispheres in tropical, subtropical and temperate zones (Tapia, 2014).

Beans are one of the oldest crops because archaeological research indicates that they were known at least 5000 years ago. It is considered that the American plant trilogy of maize, beans and squash did not exist when the bean was in the process of domestication (Mora, 2013).

It was only at the end of the last century and the beginning of the present century that the American origin of the bean was accepted, as previously the Asian continent was suggested as the centre of origin. Archaeological research led to the discovery of remains in several sites in Peru, Mexico and the United States, as well as in Chile, Ecuador, Argentina and Central America. In Peru, remains dating back 8,000 years were found in the Guitarrero caves (Tapia, 2014).

In Mexico in the Tehuacan Valley, 7000 years ago. In the southwestern region of In the United States, remains were found in the Tularosa Cave dating back to 2,300 years. These remains consist of seeds, pod fragments and plant parts collected in drought areas in both Mesoamerica and the Andes (Pacheco, 2011).

Beans are reported to be one of the first domesticated plants in the Americas, dating back 6,000 to 8,000 years ago, both in Mesoamerica and the Central Andes. Its cultivation spread to almost all pre-Hispanic populations and was one of the main food crops, along with maize, potato and cassava (Escoto, 2014).

Less than 60 years after the discovery of America in 1492, beans were widely cultivated in Western Europe. From there it spread to the rest of Europe, Iran, India, the Middle East and elsewhere in Asia and AfricaTapia (2014). It occupied other areas of the world such as the USA, Europe and tropical Africa and later North Africa and Asia, after which its cultivation became increasingly important, spreading to both hemispheres in tropical, subtropical and temperate areas, due to the adaptation of the large number of varieties and types. It is undoubtedly the most important species of the genus Phaseolus (CIAT, 2010).

About 20 species of grain legumes are used as food in significant quantities. In African, Asian and Latin American countries, grain legumes are used as a source of protein, 'poor man's meat', as they contain between 18 and 30 % protein. It is the most important legume in the diet of Central and South American countries and those located in Central and East Africa. Latin America is the largest production and consumption area in the world, with an estimated 30 % originating there (Box, 2011). In the world, beans are in great demand by the population in countries such as Brazil, India and the United States of America stand out as major producers worldwide. In the years 2007 to 2014, the world bean production was around 17,803,000 to 20,991,898 tonnes, the harvested area was 23,667,767 to 28,189,680 ha, with yields of 0.68 to 0.76 t.ha-1 respectively (CIAT, 2010).

2.1.2. World production and consumption

It is the most important legume in the diet of Central and South American countries and those in Central and East Africa. Latin America is the largest production and consumption area in the world, and it is estimated that 30 % comes from it. Africa is the second largest producer (20 to 25 %), with Uganda, Kenya, Rwanda, Tanzania and Burundi being the largest producers (Baudoin et al., 2021).

Production in 2005 was 19.91 million tonnes. The main producers were Brazil, India, Myanmar, China, Mexico, the United States and Indonesia. The United States of America was the seventh largest producer and among the 29 countries with the highest world production, two Central American countries stand out: Nicaragua (No. 20) and Guatemala (No. 29). However, the former mainly produces red beans, while the latter produces black beans (CIAT, 2014).

Binder (2007) indicates that a large part of legume production in Latin America takes place in production systems ranging from 1 to 10 hectares, located on soils of low fertility, so that around 80 percent of legume production in the region is carried out on low fertility soils. % is sown on mountain slopes.

In accordance with the multiple and independent domestication process that the crop underwent, current consumption patterns in terms of bean size and colour vary between Latin American countries. Thus, Mexico, Colombia, Ecuador, Peru and Chile have the highest demand for large beans; the opposite occurs in Central American countries, Brazil and Venezuela, which prefer small beans. In terms of bean colour, Venezuela and Guatemala are the only countries that consume almost exclusively black beans, while others prefer other colours such as: red (Colombia, Belize, Costa Rica, Salvador, Mexico, Panama); cream (Chile, Colombia, Mexico); white (Chile, Mexico, Peru) and different colours such as dark brown, light red, yellow in other countries (Ayala and Mar, 2014). Vegetables in general and beans in particular are scarcely consumed in Argentina, only 0.3 kg/inhabitant/year, while in other countries such as Brazil consumption is 20.1 kg/inhabitant/year, Mexico 12.6 kg/inhabitant/year, and Mexico 12.6 kg/inhabitant/year. kg/inhabitant/year, Paraguay 24.3 kg/inhabitant/year and Uruguay 2.3 kg/inhabitant/year. The average "per capita" consumption in Latin America is 13.3 kg/inhabitant/year (FAO, 2015).

Worldwide, most of the population receives a small portion of total dietary calories from common beans. However, for Latin America and Africa, this legume represents the third and sixth largest source of calories, respectively (Ayala and Mar, 2014).FAOSTAT (2010) highlights the use, for dry consumption, of the small, black, opaque, smooth-fronted, truncated and ovoid type in eastern Venezuela. Others agree that the origin of this preference goes back to long ago when the dispersions of black beans from Guatemala advanced towards the interior of Teuantepeque and the Yucatan Peninsula, from where the Caribbean tribes transported them to Cuba and from there to Venezuela and spread to Brazil. This involves the Caribbean island in their dispersion on the continent.

2.1.3. Revenues at worldwide

Baudoin et al. (2021) note that under favourable conditions and with the use of high-yielding varieties, up to approximately 5 $kg.ha^{-1}$; however, the reality is very far away, especially for smallholders, whose yields are usually much lower, with 2.8 $t.ha^{-1}$ reported for the USA; Africa (Moderate) 0.8 $t.ha^{-1}$; South Central America 1.8 $t.ha^{-1}$; Oceania, 1.3 $t.ha^{-1}$ and Asia 1.2 $t.ha^{-1}$. Lee (2001) states that the yield potential of beans is greater than 3$kg.ha^{-1}$, but averages in Latin America are low, around 600 $kg.ha^{-1}$. This is influenced by several factors, including: variable climatic conditions, soil nutritional problems, socio-cultural and economic conditions of farmers, low use of technologies, pests and diseases, among others. There is a significant difference in yields obtained in traditional cropping systems with those obtained in Research Stations when the best cultivars and optimal growing conditions are provided (Baudoin et al., 2021).

2.1.4. Chemical composition and nutritional value .

Beans (Phaseolus vulgaris L) are an important source of protein; they contain about 20 % of highly digestible proteins, composed of amino acids essential for human metabolism, such as isoleusine, leusine, phenylalanine, methionine and tryptophan. In addition, it can also be considered a food with high energy value, as it contains between 45 and 70 % of total carbohydrates. It also provides significant amounts of minerals (Fandiño, 2009). Therefore, beans (Phaseolus vulgaris L.) have a high nutritional value in comparison with other foods. Table1.

Table 1. Nutritional value of beans (Phaseolus vulgaris L.).

FOOD	Água	Calorie (p/100g)	Protein (%)	Grassa (%)	carbohydrates
Bean	11	341	22.1	1.7	61,4
Soybeans	8	354	38.0	18.0	31.3
Rice	13	360	6.7	0,7	78,9
Maize	12	360	9.3	4.0	73,5
Wheat	13	360	6.7	0	78,9
Cassava flour	11	338	3.8	0,6	81,5
Eggs	74	158	13.0	11.0	0,7
milk powder complete	2.5	498	27,5	2.6	28.0
Beef	67	198	19.0	13.0	0
Fish	65	75	16.4	0,5	0

Source: Fandiño, (2009) and cited by Pequenino (2018).

It also contributes positive effects in the prevention and treatment of cardiovascular diseases, diabetes and cancer, both for its contribution of micronutrients (particularly folic acid and magnesium) and for its high content in fibre, amino acids, sulphur, tannins, phytoestrogens and non-essential amino acids (Rodríguez, 2006).

It is an energy food (as it provides a lot of calories), which provides protein and, to a certain extent, also serves as a regulator, due to its vitamin and mineral content. If consumed together with a vegetable rich in methionine, although low in lysine, the two complement each other, resulting in a protein with good biological value (Valdivia, 2010 cited by Tapia, 2014).

2.2. Taxonomic classification of the bean crop.

According to Socorro et al., (2008), this species is classified as:

Table 2. Taxonomic classification of the bean crop.

Kingdom:	Plantae.
Division:	Magnoliophyte.
Class:	Magnoliopsida.
Subclass:	Rosidae.
Order:	Fabales.
Family:	Fabaceae.
Subfamily:	Faboideae.
Tribe:	Phases.
Subtributing:	Faseolinae.
Gender:	Faseolo.
Species:	Phaseolus vulgaris L.

Fifty-five species are reported, of which four are cultivated: P. vulgaris L.; P. coccineus L.; P. lunatus L. and P. acutifolius A.

Common name: Bean

2.3. Characteristics morphological characteristics

✓ Source: Hernández (2012), in the first stage of development, the root system is formed by the radicle of the embryo, which then becomes the main or primary root. After a few days of radicle emergence, it is possible to observe the secondary roots, which develop especially in the upper part or neck of the main root. Tertiary roots and other subdivisions, such as absorbing hairs, develop on the secondary roots, which are also found at all root growth points. In general, the root system is superficial, as the greatest volume of roots is found in the first 20 centimetres of soil depth. Although the taproot is generally distinguishable, the root system tends to be fibrous in some cases, but with wide variation even within the same variety. As a member of Subfamily Papilionoideae, Phaseolus vulgaris L. has nodules distributed on the lateral roots of the upper and central part of the root system. These nodules are colonised by bacteria of the genus Rhizobium, which fix atmospheric nitrogen, helping to meet the plant's requirements for this element.The composition of the bean root system and its size depend on soil characteristics such as structure, porosity, aeration, moisture holding capacity, temperature and nutrient content.

✓ Source: Tapia (2014), The stem can be identified as the central axis of the plant, which is formed by the succession of nodes and between nodes. It originates from the apical meristem of the seed embryo. From germination and in the early stages of plant development, this meristem has a large apex and in the process of development generates nodes. A node is the point of insertion of leaves or cotyledons into the stem. The stem is herbaceous and has a cylindrical section, due to small wrinkles in the epidermis.

The stem is the result of a dynamic process of active construction from its earliest stages of growth by a cluster of cells located at the apex, called the apical meristem. This construction process also includes the formation of other organs within and between the nodes.

The stem is generally larger in diameter than the branches and may be erect, semi-erect or creeping, depending on the growth habit of the variety. There is variation in the pigmentation of the stem, so that three basic colour derivations can be found: green, brown and red.

✓ Leaves: Araya (2010), Bean leaves are of two types, simple and compound, and are inserted at the nodes of the stem and branches. Primary leaves are simple and appear at the second node of the stem, are formed on the seed during embryogenesis and fall off before the plant is fully developed.

Trifoliate compound leaves are the most typical of beans, with three leaflets, petiole and rachis. At the insertion of the trifoliate leaves there is a pair of triangular-shaped stipules which are always visible.

Inflorescence

According to researcher Camellón (2012), inflorescences can be terminal or axillary. From the botanical point of view, they are considered as clusters of clusters, i.e. a main cluster composed of secondary clusters, which originate from a complex of three buds (floral triad) that are are found in the axils formed by the primary bracts and the rachis. Three main components can be distinguished in the inflorescence: the inflorescence axis formed by the peduncle and rachis, the primary bracts and the flower buds.

✓ Flower: Tapia (2014), The bean flower is a typical papilionaceous flower. In the development process of this flower, two stages can be distinguished, the flower bud and the fully open flower.

The flower bud, whether it originates from the insertions of a raceme or from the floral development of the buds of an axil in its initial stage, is surrounded by bractlets which are oval or round in shape. In its final stage, the corolla stands out, which is still closed and the bractéoles only cover the stem. When the phenomenon occurs before the flower opens.

The floral morphology of the bean favours the self-pollination mechanism, as the anthers are at the same level as the stigma. When pollen is shed (anthesis), it falls directly on the stigma.

✓ Fruit: Camellon (2012), The fruit is a pod with two valves coming from the compressed ovary. As the fruit is a pod, this species is classified as a legume.

The sheaths can be of different colours, uniform or grated, depending on the variety. Two sutures appear at a union: the dorsal suture, called the placental suture, and the ventral suture. The ovules, which are the future seeds, alternate in the placental suture.

✓ Seed: Tapia (2014), The seed has no albumen, so nutritional reserves are concentrated in the cotyledons. It can have various shapes: ovoid, cylindrical, kidney-shaped.

2.4. Growth habit of bean

For CIAT (2014), according to their growth habit, they are classified as follows:

✓ Type I. They generally have few nodes (5 to 10), end in an inflorescence, stand erect, tend to be large-seeded, precocious, with a short flowering period, low yield potential (although this can be compensated for by higher plant density), more uniform maturity, strong and thick stem, relatively long pod length, and are generally soft cooking and thick juicy. These varieties respond well in furrows 30 to 60 cm wide.

✓ Type II. They are erect, have a small guide on the main stem and the branches do not produce guides, have greater yield potential and a greater number of nodes 11 to 14 than Type I, are usually of new pods and seeds, intermediate to delayed life cycle, respond well in furrows 40 to 70 cm wide.

✓ Type III. They have high yield potential, higher number of nodes and branches of 12 to 16, of various colours and grain sizes, their vegetative cycle is intermediate to delayed. These varieties respond well in 60 to 70 cm wide furrows.

✓ Type IV. They are usually ascending, multi-coloured, have high yield potential, with 14 to 18 nodes and usually have a delayed cycle of more than 120 days. These varieties respond well in 70 to 80 cm wide furrows.

✓ Type V. Delayed cycle (120 to 160 days), with 16 to 30 nodes, high yield potential, various seed colours and sizes, generally

Table 3. Developmental stages of the bean plant.

Steps	Code	Name
Vegetative	V0	Germination.
	V1	Emergency.
	V2	Primary leaves.
	V3	First sheet trifoliate.
	V4	Third sheet trifoliate.
Reproductive	R5	Pre-flowering.
	R6	Flowering.
	R7	Pod formation.
	R8	Full of pods.
	R9	Maturation.

Source: Camellón, (2012).

Phases of the vegetative stage: in this phase the stages of germination, emergence and leaf formation occur. Stage V0 (Germination). The seed absorbs water and the phenomena of cell division and biochemical reactions that release nutrients from the cotyledons take place. The radicle soon emerges, which then develops into a root. primary when secondary roots appear on top of it; the stem also grows and the cotyledons are at ground level.

Stage V1 (Emergence). Starts when the cotyledons appear at ground level. The stem straightens and continues to grow, the cotyledons begin to separate and soon the primary leaves develop.

Stage V2 (Primary leaves). Starts when the primary leaves of the plant unfold. In cultivation, this stage is considered to begin when 50 % of the plants show this characteristic. At this stage the rapid vegetative development of the plant begins, during which the stem, branches and trifoliate leaves will form. The cotyledons lose their shape by shrivelling and arching.

Stage V3 (First trifoliate leaf). Starts when the plant has the first trifoliate leaf fully open. In cultivation, this stage begins when 50 % of the plants have unfolded the first trifoliate leaf.

Stage V4 (Third trifoliate leaf). This stage begins when the third trifoliate leaf unfolds. In cultivation, this stage begins when 50 % of the plants show this characteristic. From this stage onwards, differences can be clearly observed in some vegetative structures such as the stem, branches and trifoliate leaves which develop from the triads of buds. The first branch usually starts to develop when the plant enters the V3 stage.

According to Camellón (2012), the stages of the reproductive phase include pre-flowering, flowering, pod formation, seed growth and ripening, these stages are:

► Stage R5 (Pre-flowering). The R5 stage begins when the first bud or flower cluster appears. In cultivation, this stage is considered to begin when 50 % of the plants show this characteristic. In varieties with determinate growth, the development of flower buds is noticed in the last node of the stem or branch; while in varieties with indeterminate growth, flower clusters are observed in the lower nodes.

► Stage R6 (Flowering). Stage R6 begins when the plant shows the first open flower in cultivation, when 50 % of the plants show this characteristic. The first open flower corresponds to the first flower bud to appear. In varieties with a determinate growth habit, flowering starts at the last node of the stem or branches and continues downwards at the lower nodes. On the other hand, in indeterminate-growing varieties, flowering starts at the bottom of the stem and continues upwards. Once the flower has been fertilised and is open, the corolla fades and the pod begins to grow.

► Step R7 (pod formation). In the plant, this stage begins when the first pod appears with the flower corolla hanging down or detached, and under cultivation conditions when 50 % of the plants show this characteristic. During the first 10 to 15 days after flowering, there is mainly longitudinal pod growth and little seed growth. When the pods reach their final size and maximum weight, seed growth begins.

► Stage R8 (Seedling growth). In cultivation, the R8 stage begins when 50 % of the plants start to fill the first pod. Active seed growth then begins. At the end of this stage, the grains lose their green colour and begin to acquire the characteristics of the variety. In some varieties, the colours of the pods begin to pigment, which usually occurs after seed pigmentation begins.

► Stage R9 (Ripening). This stage is the last on the developmental scale, as the crop matures. It is characterised by the ripening and drying of the pods. This stage begins in the crop when 50 % of the plants have at least one pod, discolouration and drying begins. As the pods dry, they lose their pigmentation; the water content of the seeds drops to between 15 and 20 %, at which point they attain their typical colour. This is the end of the biological cycle of the plant and it is ready for harvesting.

2.5. Agronomic work required by cultivation.

► Area selection

Preferably on flat soils and avoiding low areas with danger of waterlogging, as well as good internal and surface drainage and an optimum pH: 5.0 to 8.1. Without the presence of obstacles (Escoto, 2014).

► Soil preparation

With this crop, the cultivation aims to create a suitable layer for the germination of the seeds, with the objective of ensuring that the plant obtains optimal conditions for its growth and development. It is necessary that the The number of ploughing operations should not be excessive so as not to destroy the granular composition and avoid compaction (Escoto, 2014).

Between the start of preparation and sowing, a period of time should elapse to allow weed and crop residues from previous harvests to compose themselves. In addition, two to three germinations of sprouting weed seeds should be removed

(Araya, 2010).

In mechanised sowing it is necessary to remove obstacles and leveling. Minimum tillage is carried out without inverting the prism when animal traction is used. The depth of the break and the rest of the work must not be less than 25 cm. The soil should be subsoiled every three to four years to a depth of no less than 40 cm (Ramos, 2014).

► Sowing

The most favourable sowing season for the cultivation of Phaseolus vulgaris L. is the one that allows harvesting to coincide with the period of little or no rainfall, to avoid damaging the grains due to excess humidity, as well as offering the climatic conditions for the good development of the crop (Ramos, 2014).

► Sowing distance

This will depend on the variety, seed size and germination. The sowing distance for varieties with a determinate upright creeping growth habit types III and II is 45 to 70 cm between rows and 5 to 7 cm between plants. Varieties with determinate growth habit type I should be sown in double furrows of 30 to 60 cm and 7 cm between plants. The sowing depth should be between 3 and 5 cm (Araya, 2010).

► Fertilisation

Beans are grown in soils with highly variable physical and chemical conditions, nutritional deficiencies can affect the development and yield of the crop. The information available from various authors is variable with different varieties and populations (250 to 300 thousand plants.ha^{-1}), the average nutrient uptake is 133.8 - 16.0 - 116.6 kg.ha^{-1} and an average extraction and export in the seed is 32.2 - 5.4 - 17.2 kg.t^{-1} of nitrogen, phosphorus and potassium respectively, to achieve the yield potential of the varieties, complete fertilizer should be applied at the bottom of the furrow at the time of planting (Araya, 2010).

► Irrigation

Beans require approximately 9 irrigations (with a clear standard of 250 m3.ha-1) during the whole crop cycle, regardless of the variety and soil type, the soil must remain at 80 % of the field capacity. Beans have 4 critical stages: germination, flowering, seed formation and growth, where water must not be lacking so that yields are not affected (Ramos, 2014).

► Harvest

Harvesting must be done at the beginning of the technical maturity stage, when the grain has between 15 and 18 % moisture. It can be done mechanised or manually (Araya, 2010).

2.6. Conditions edaphoclimatic

Cabrera (2000), states that the bean thrives in cold and warm climates, and has climbing and dwarf varieties. It is cultivated in low-saline soils, with a medium level of rainfall. The same author mentions that this plant is grown in places where the heat of the sun reaches the stem of the plant. Although it tolerates a wide range of soils, the most suitable are light, siliceous-loamy soils with good drainage and rich in organic matter. In strongly clayey, very calcareous and saline soils with little vegetation, it is very sensitive to excess humidity, so an excess of irrigation can be enough to damage the crop, leaving the plant yellowish and small. Ayala and Mar (2014), mention that the cultivation of beans requires fertile soils, with good organic matter content, the most suitable soil textures are: moderately heavy, good aeration and drainage, as it is a crop that does not tolerate compact soils with poor aeration and water accumulation.

Hernández (2012), indicates that optimum pH values range between 6.0 and 7.5, although in sandy soils it does well with values of up to 8.5. If the soil is light and sandy, add plenty of moist peat, fertiliser or tanned manure. If the drainage is not good, a mound will form and seeds will form on top. If the soil is very acidic, add lime. Climatic factors such as temperature and light are not easy to control. but it is possible to control them, using cultural practices, such as the sowing at appropriate times, so that the crop has favourable conditions (Hernández, 2012). The Centro de Investigaciones Agrícolas Tropicales (CIAT), (2010), states that water is essential for crop development and yield. There are lines and varieties that show good tolerance to water deficiencies, giving acceptable yields under these conditions, a tolerance that can be supported by the greater capacity to extract water from the deep layers of the soil.

Arias et al. (2009) state that the most important role of light is in photosynthesis, but it also affects plant phenology and morphology. Beans are a short-day species, long days tend to cause delays in flowering and maturity. For each additional hour of light per day, ripening can be delayed by two to six days.

Meteorological variables can be highly variable, both in the way they manifest themselves and in their magnitudes, which makes it difficult to manage and use them appropriately. Therefore, it is necessary, taking into account the plant's requirements in terms of climatic conditions, to be able to combine them with their impact over time, so that they act favourably. To do this, it is necessary to know the behaviour of the climate in the area, to adjust it to the sowing season, so that adverse climatic conditions do not occur (Socorro et al., 2008 cited by Camellón, 2012). The way in which meteorological variables are presented requires the establishment of forecasts based on accumulated historical data, especially the incidence of rainfall, high or low temperatures and winds (Escoto, 2014).

► Temperature

Temperature influences bean cultivation throughout its cycle Socorro et al. (2008) state that the optimum temperature for plant development is 24 to 25 ºC, although for each stage of development there are optimum temperature ranges, maximum and minimum values. The optimum temperature range for the cultivation of beans during germination is a minimum of 8 ºC, the optimum is 24 to 25 ºC and the maximum is 30 ºC, for flowering the minimum temperature is 15 ºC, the optimum is 22 to 25 ºC and the maximum is 30 to 35 ºC, for ripening the minimum is 18 ºC, the optimum is 22 to 26 ºC and the maximum temperature is 26 to 30 ºC. During the vegetative development of the plant, and up to temperature regulates the processes of respiration, photosynthesis and nutrient uptake.

► Socorro et al., (2008) and called by Escoto, (2014)

The relative humidity surrounding the bean plant can directly or indirectly influence its development, directly on the physiological functions performed by the plant, such as transpiration, respiration and photosynthesis. A relative humidity of 70 % is considered adequate for the good development of the bean plant. Soil moisture can also have a positive or negative influence on crop development. The bean's water requirements are low, so that the accumulation of excess moisture in the soil, as a result of heavy and poorly distributed rainfall or excessive irrigation, especially in soils with drainage difficulties, causes significant alterations in the development of the root system and in the rest of the plant. It should be remembered that the root system of the bean is underdeveloped and, therefore, needs to be able to achieve the greatest possible development. Therefore, soils that, due to their characteristics, retain a lot of moisture are not the most suitable for this crop. However, when sowing beans in the dry months of the year, irrigation becomes essential to guarantee the crop's water requirements.

► Luz.Socorro et al., (2008) and called by Camellón, (2012)

Socorro et al. (2008) state that light is a meteorological variable with particular implications for bean productivity. Its action is known to condition plant growth and development by providing the energy source for the photochemical phenomena that regulate the plant's physiological processes. However, it is difficult to determine its effect on the plant in isolation due to its close connection with other factors such as temperature and humidity. The action of light on the bean plant can be estimated on the basis of the amount of light received (which depends directly on the length of the day and the effective radiation) as well as its quality (which depends on the type of radiation).

Bean is a short-day crop, so flowering is favoured by photoperiods of less than 12 hours with long periods of darkness, which occurs in Angola from May onwards. On the other hand, it is possible to observe tendencies in the plant towards a continuous growth with the appearance of a fickle when growing under prolonged lighting conditions, causing limitations in the reproductive phase. This circumstance means that flowering is progressively reduced if the cycle is lengthened and there is more leaf development. However, the differences between maximum and minimum day length at our latitude do not seem to be pronounced enough to completely inhibit flowering, although they can affect it.

► winds

Winds have a negative influence when they occur at high speeds, as the plant has a large leaf volume, increases the transpiration rate and, therefore, cannot always compensate well for the desiccation that occurs in the leaves due to the limitations of the root system (Socorro et al., 2008).

2.7. Performance characteristics and their components .

The bean yield is composed of: number of inflorescences (bunches) per plant, number of pods per bunch, number of seeds per pod, the weight of the seeds in turn is determined by their components, length and width. In legumes, the properties of the pods play an important role. The heritability of yield in general is low or similar to that of other yield components (Camellón, 2012).

2.8. Main bean pests. Characteristics (Martínez et al., 2007)

✓ Empoascapp (leafhopping)

It has six white spots on its head and thorax. Attacked plants turn a deep green colour and their leaves curl up sharply.

✓ Bemisiaspp (whitefly)

The whitefly is one of the most important pests worldwide. The economic importance of this insect is due to its wide geographical distribution in tropical, subtropical and temperate areas of the world, the large number of cultivated species it affects and the large number of wild plants it attacks. The adults and nymphs of this insect suck sap from the phloem. This is a direct damage that reduces income. The production of sugary secretions by adults and nymphs indirectly affects production because it favours the development of fungi that interfere with photosynthesis.

The larvae have rudimentary legs which serve to attach themselves to the underside of leaves where they remain until the pupal stage. The adult is white in colour and appears in the first few days after sowing the crop, increasing until the flowering stage. It is the main vector of the Golden Mosaic Virus in beans, a disease that considerably reduces yields.

✓ Polyphago tarsonemuslatus (white mite)

Symptoms are most visible on young leaves, which turn purple on the back and sometimes dark yellow. The first symptoms are observed as a rolling of the nerves in the leaves and apical shoots, with curvatures in the more developed leaves. In more advanced attacks, dwarfing and intense green colouring of the plants, flower abortion and a general stiffening of the vegetative organs of the plants occur. It is a very small mite and can only be observed with the aid of a good magnifying glass or stereoscope (Travuco et al., 2016).

In the adult stage they are light green in colour, the underside of the leaves turn purple, when the condition is older the leaves may turn dark yellow.

✓ Diabroticabalteata (common chrysomelid)

The larvae feed exclusively on roots and tubers, reducing the vigour of the plant. Their production has a direct impact on the quality of produce in certain crops such as sweet potatoes. Adults are particularly damaging to small plants and completely eat the cotyledons. In adult plants, although direct damage is limited, they can cause serious problems, especially through disease transmission (López, 2014).

✓ Cerotomaruficornis Olivier (common bean chrysomelid). Synonym: Andrectorruficornis

Adults feed on foliage, puncture leaves and cause a reduction in leaf area. The greatest effects are seen in the early phenological stages of the crop, although it affects almost the entire crop cycle. This pest is a vector of the yellow spotted slug (BYSV) and cowpea mosaic virus (CPMV) (López, 2014).

✓ Uromycesphasioli (kidney bean)

Small, white to reddish-brown spots, varying in size from 1 to 2 mm, with a large number of spurs, appear on leaves, sheaths and sometimes on stems and branches.

✓ Colletotrichum lindemuthianum (Anthracnose)

On the leaves, initially on the underside, reddish-brown to black lesions are found along the veins which may develop into cankers, which may coalesce and form neurotic areas at the edge. Petioles, branches, stems, cotyledons and pods may be attacked. Lesions are small, depressed, brown, often covered by masses of spurs.

✓ Xanthomonas campestripv. Phasioli (common bacteriosis)

On the leaves, small wet spots are found on the back of the leaves, which enlarge irregularly and coalesce. The infected parts become flaccid, brown in colour and may be surrounded by a yellow thread. On the pods, the damp spots gradually become dark to reddish in colour and are slightly depressed.

✓ RhizoctoniasolaniKuhn (asexual stage), Thanatephorus cucumeris (Frank) (sexual stage).

T. cucumeris attacks mainly leaves, stems, twigs and pods at any stage of development; it does not cause damage to roots. The first symptoms appearing on leaves are usually caused by splashing of sclerotia and mycelium (asexual stage). Another type of lesion, in this case caused by sporulating basidia (sexual stage).

✓ Common bean mosaic (Common bean mosaic necrosis virus, Common bean mosaic necrosis virus, Common bean mosaic necrosis virus, Common bean mosaic necrosis virus).

bean - BCMNV)

The main symptom of common mosaic is root necrosis (black root), although this may vary with environmental temperatures and species genotype. In general, it appears as a green mosaic (light green spots alternating with dark green) and turns the leaves upside down. Deformed leaves may also appear on the same plant.

In general, the development of plants with few and small pods, which are covered with small dark green spots, is reduced. These symptoms can easily be confused with Cucumber Mosaic (CMV), but they differ in that they are transitive in CMV and permanent in BCMNV.

✓ Golden bean mosaic (BGMV)

Transmitting agent: Whitefly (Bemisiaspp)

Young plants show symptoms on the first trifoliate leaves, with the veins turning light yellow. This process usually begins in the middle of the leaf near the tip.

After three to five days, this vein extends to cover a large part of the leaf, contrasting with the dark green interveinal areas.

✓ Yellow mosaic (Bean yellow mosaic virus BYMV)

Symptoms of this disease may vary depending on the species, virus strain and phenology of the plant at the time of infestation, as well as environmental factors. They usually consist of leaf and rib colours, a mosaic appearance, as well as leaf deformations and stunting of small plants.

✓ Erysiphebetae (powdery mildew)

Symptoms appear first on the leaves, starting with the oldest leaves and later on the pods and stems. They appear as white, powdery-looking spots on the leaf surface. The tissue beneath these white spots changes colour due to the action of the fungus, turning purplish-brown. In case of severe attacks, the entire leaf surface may be affected. The tissues become necrotic and the leaves undergo a premature ageing process until they die.

According to Vázquez (2008), the parts of the plant where the attack by pest are shown in tables 4 and 5.

Table 4. Parts of the plant where attack by the main pests is concentrated.

Pests	Provenance	Sheets	Fruit
Bean Rust		unknown	unknown
anthracnose	unknown	unknown	unknown
downy mildew dusty		unknown	
Chrysomelids		unknown	
white fly		unknown	
jumping leaves		unknown	
white mite		unknown	

Source: Vázquez, (2008) cited by Pequenino, (2018).

Table 5. Predominant weeds in the bean crop.

scientific name	cycle vegetative	Form of propagation	Features
Amarantus dubius Mart.	Perennial	For seeds	Strongly aggressive and very competitive.
Cyperusrotundus L.	Perennial	For seeds	Causes damage to the first states.
Echinocloa colona L.	Annual	For seeds	It is a powerful invader.
Euphorbia heterophylla L	Annual	For seeds	It is very aggressive and strong.
Portulaca olearaceae L.	Annual	By seed and portions of the stem.	Easy propagation
Sorghum pense Sibthorp.	Perennial	By seed and portions of the stem.	It is more harmful.

Source: Franco et al., (2014).

Franco et al., (2014) include, in addition to pests, a number of undesirable plants that cause damage to the bean crop, including those mentioned in Table 4.

2.9. Control measures pests

Pest control can be preventive or curative; the former takes place before pathogen attack; and the latter is when the crop is already affected (Almaguel, 2002). Taking into account the damage that pests and diseases can cause to crop yields and product quality, and in the deterioration and growth of the flora in the entity, this phenomenon should be considered as low intensity, when it affects an area and a certain type of crop, which can be controlled almost immediately; medium intensity, when the risk and damage caused are unacceptable, significantly affecting production; and high intensity, when its effects are highly destructive and seriously alter the environment, generating a gradual and imminent risk situation for the population (García et al., 2009). Pest management refers to reducing pests to levels acceptable to the farmer, but does not necessarily imply their absolute elimination. Although minimal use of pesticides is suggested, this alternative is not ruled out, when the situation warrants it. This means that there are situations in which the application of a chemical is the most logical and valuable measure, as long as this decision is taken considering ecological, socio-economic and environmental criteria in general (Almaguel, 2002).

Many organisms have enormous reproductive power and would develop rapidly if there were no natural control mechanism to keep them at a certain level without direct human intervention. When designing management measures it is always important to know the natural regulation of pests (natural control), as this is a very important factor and often determines whether or not an organism reaches pest status (Hernández, 2012). Natural control consists of the collective action of physical and biotic environmental factors that maintain the pest at a certain oscillating level over a period of time. Within the components of natural control, climatic factors (rain, temperature, wind) and natural enemies (parasites, predators and pathogens) play an important role. Therefore, our applied control actions should aim to take advantage of these pest control factors (Hernández, 2012).

2.10 Organic fertilisers on crops

Yalta (2001) in the work "Effect of Mulch with incorporation of chicken manure on the yield of cabbage crop (Brassicaoleracea, Var. Capitata alvorada L.), concludes that the most important yield obtained among the mulches used (guava leaves, rice cone, kudzu leaves, sawdust), was with guava leaves; with this treatment a head/ha yield of 35.33 t.ha-1 was obtained. García (2013) in the thesis "Dosage of soft bed subsidy (sow + rice cone) and its effect on the yield of cabbage leaves (Brassicaoleracea L.), Var. Tropical Delight in Fundo Zungarococha", concludes that the maximum yield obtained from

Brassicaoleracea L., Var. Tropical Delight "cabbage", was in treatment T4 (9 kg of leaf litter/m^2) with 36, 720 $t.ha^{-1}$ and the minimum was treatment T1 (3 kg of leaf litter/m^2) with 29, 940 t. .ha $.^{-1}$

Torres (2012), in the thesis "Sowing and cover methods, their influence on agronomic characteristics and yield in Brassicaoleracea L. "cabbage", Var. Capitata alvarada, Tropical Delight hybrid in Nina Rumi - San Juan Batista, concludes that the T2 treatment (single row with organic cover) obtained a cabbage weight of 1.77 kg and the T1 treatment (single row + synthetic cover) obtained a yield of 1.53 kg of cabbage weight.

Zagaceta (2001), in the thesis "Effect of sowing distance in the cultivation of Brassicaoleracea Kale, Var. Capitata alvarada L. with applications of poultry manure", concludes that the distance of 60 cm between rows and 30 cm between plants gives the best yield, with 42,570 $t.ha^{-1}$.

Chujutalli (2008), in the work "Influence of the application of Biol and its effects on the yield of Brassicaoleracea L. "cabbage, Var. Tropical Delight in Zungarococha", concludes that the maximum yield obtained in Brassicaoleracea L. Var. "Cabbage" Tropical Delight, was in treatment T1 (Biol with a frequency of 3 days), which obtained a yield of 40,400 t.ha $.^{-1}$ Jaramillo and Leyva (2002) mention that during the decomposition process of organic matter, organic and inorganic acids are formed, which influence the acidity of soils; at the same time, they show that sulphuric and nitric acids are formed not only by the process of organic degradation, but also by microbial action on certain fertilisers such as sulphur and ammonium sulphate, specifically the latter. Camagro (2005), states that, in the course of its development, a plant consumes a certain amount of certain elements which varies according to the species and which must be restored in the form of inputs, depending on the nature of the soil and the needs of the crop. Organic fertilisers are also considered as amendments, as they correct physical properties; they provide considerable quantities of nutritional elements, producing chemical changes in the soil. IMPULSEMILLAS (2007) mentions that soil organic matter is derived from the remains of dead plants and animals and dead organisms in the soil. Thus, the relative compounds are those that were part of living networks, carbohydrates and related substances, lipids and proteins. In general, these compounds are oxidised to the end or converted into humus.

2.10.1. Fertilisation and yield of crops.

Babilônia and Reátegui (1994) plant that initially requires the use of 5 kg. of barnyard guano (galinaza) per m2, mix well and let stand for a week, time elapsed and 30 hours before planting, you must add a complete fertilizer (NPK) at the rate of 54 g of Urea, 27 g of Phosphate. triple, 50 g of Potassium Chloride, but 10 g. of lime (live or prepared) and 5 g of Magboro per m2.

Ministry of Agriculture (1997) indicates that the yield of cabbage leaves in Peru in order of merit corresponds to the following regions: Lima with 19,572 $kg.ha^{-1}$; Arequipa with 16,610 $t.ha^{-1}$ and Andrés Avelino Cáceres (Huánuco, Pasco, Junín) with 14,318 $t.ha^{-1}$, the Loreto region accumulates 4,318 $t.ha^{-1}$; Furthermore, the same publication mentions that the area cultivated this year in Lima with 1181 ha, Chavín with 552 ha, Andrés Avelino Cáceres with 518 ha and Arequipa with 433 ha.

In this regard, Pinchi (1999) states that using laying hens at a rate of 5 kg/m^2 , 8 days before transplanting and applying NPK after 30 days at a rate of 7.5 g of mixture/plant, 19,080.00 $kg.ha^{-1}$. were obtained with Tropical Delight income.

Jercy (1996) states that the nitrogen (N) requirement is 100 to 225 $t.ha^{-1}$, distributed in one to three applications, banded on both sides of the furrow, before the beginning of head formation. Low doses are recommended when cabbage has been planted after a heavily fertilised crop, on clay soils or when environmental conditions favour accelerated crop growth. The same author refers to this in the case of phosphorus (P). On soils poor in this nutrient (less than 15 ppm) 225-280 t.ha^{-1} of P2O5 are recommended. On medium soils (15-30 ppm) 170-225 kg.ha^{-1} are applied in the same way. For soils with good phosphorus content (+30 ppm) a fertilisation of no more than 90 kg.ha^{-1} can be used. And in the case of potassium (K). In soils that require the application of this nutrient, it is advisable to use a dose of 110-220 t.ha^{-1} of K2O and the application should be made by volley to incorporate it into the soil before spreading the beds.

2.10.2. Organic fertilisers. livestock manure

On the loan Galvão (2013) states that manure is a very important fertiliser and could successfully address the until recently problem of nitrogen, which is the element that is most lost during burning.

Gratelly (1992) also states that manure improves soil aggregation, makes the soil more absorbent of rainwater, improves drainage and forms a surface layer of humus that reduces the erosive action of rainfall.

On the other hand, Gutiérrez (1994) states that much attention should be paid to the combined use of manures and organic fertilisers to increase agricultural production and maintain soil fertility. In the same way, he states that manure is mainly used in pastures, gardens, orchards, but there is no doubt that, if it is enriched with mineral fertilisers, it could be used to grow cereals and tubers intensively, in addition to the advantage of the action. of fresh organic matter is the increase of humus in the soil.

On the other hand, Valadez (1996), the contributions of manure, independently of its beneficial action as an organic amendment, make available to the crop fertilising elements which are released slowly and which the crops take advantage of in successive years by the animal species. where they come from and also by the degree of humidity, processing time, the form in which it is presented. Others who referred to dieresis were Arteaga and Pumalpa (1992) who state that the nutrient content of manure tends to fluctuate widely depending on the type of animal it comes from, the forage it receives and the maintenance it provides.

Syngenta (2007) also indicates that manure made from livestock dung is the most important of the organic fertilisers, because it that all organic substances in the manure are transformed into humus and this enhances the physical properties of the soil, making it soft and hydroscopic.

FAO (1979) indicates that studies in Asian countries report that livestock manure is a good fertiliser and is used directly on intensively cultivated areas and horticultural crops. It also increases crop yields and improves soil structure. In the laboratory, manure was found to reduce the concentration of Al and Fe ions in the revised solution, possibly due to the ratio between these compounds.

Hernandez and Chailloux (2001) mention that the use of manure, grasses and legumes in rotations is also advantageous in the control of diseases and nematodes; this is because it increases water penetration through plant residues and also improves soil structure so that there are no obstructions to drainage. The widespread use of animal manure and other organic materials will undoubtedly contribute to the absence of elemental deficiencies in many countries, not to mention the preservation of soil structure over many years of cultivation.

2.10.3. Organic fertilisers. manure from goat manure

It is one of the most active fertilisers and warmer than other fertilisers, which makes it advantageous for strong and cold soils, which it favours by dissecting them. Due to its nature and the amount used in its formation, it has a great influence on its action on the soil (Borrero, 2005).

Its effect is more immediate. Crops fertilised with goat manure are very prone to become addictive; this depends more on the condition of the animal. For example, barley fertilised with castrated goat manure produces less starch and its grains germinate unevenly (Cossanga, 2000).

According to studies by Speck (2012), land fertilised with manure from castrated goats is generally recommended; in this way, goat droppings are less exposed to wetting, and the volatile particles that are released are fixed in the soil instead of being lost.

According to Flores (2012), organic matter from goats is a good way to provide fields as well as grassland with a fertiliser that acts strongly and quickly and whose effects are particularly effective on cereals and other economic crops. importance. Fertilisation can also be done long after sowing, if the soil is not yet ready for sowing. is not too compact and not too wet. This is an excellent way to strengthen young plants when they are weak and sick. On sandy soil, the herd not only works through the manure, but also through the footprints, which will ensure more body to the soil. The usefulness of this practice is so well recognised in the Ardennes that people are no longer wary of grazing cattle on sown land, when circumstances permit. When the herd works on land that has not yet been sown, it is necessary to cover the goats' manure with a shallow plough without delay. The hotter it is, the more haste is needed (Tortosa et al., 2012).

Agrochemical characterisation of goat manure.

Parameters	Value
Humidity (%)	38,5
pH	8.5
Electrical conductivity (dS m-1)	11.33
Organic matter (%)	45,6
Total nitrogen (TN) (g.kg)-1	17.7
C/N ratio	14.3
Phosphorus (P) (g.kg)-1	2.2
Potassium (K) (g.kg)-1	16.6
Calcium (Ca) (g.kg)-1	11.9
Magnesium (Mg) (g.kg)-1	18.7

Source: Tortosa et al., (2012)

III. MATERIALS AND METHODS

3.1.Location of the experiment

The experimental study was conducted in the months between February and June 2023, at the Cristiano Shafodino mine in the locality of Caxilla III. Belonging to the municipality of Cuanhama, located in the province of Ondjiva. The research consisted of evaluating bean varieties (Phaseolus vulgaris, L.) waiting coinado, planted in a semi-arid soil, the main characteristics of the soils of this orchard are representative of the areas in which this crop is developed in the province.

✓ Beans await brother-in-law

Life cycle: 80 days; growth habit: Type II; pods per plant: 23; mass of 100 seeds: 40 g; plant height: 65 cm; colour of grain: cream pink; grain size: medium; yield in $kg.ha^{-1}$: 1300.38; number of pods per plant: 23; between nodes: 12. The experimental area is located at an altitude of 1100 metres above mean sea level, according to sheet no. 422 of Table 1:100,000 of the Instituto Geofigurae Cartofigurade Angola, (IGCA) 2008; Carta Natal/Ciudades/Angola, 2012). Beans are one of the most loved foods by Angolans. Beans, in their various forms and types, are a super-powerful legume for our diet.

Benefitsbeans awaits brother-in-law

- B vitamins, especially vitamins B1, B2, B3 and B9;
- Lysine, an amino acid that is not produced by the body and must be included in the diet. It is one of the main responsible for bone growth during puberty;
- In addition to the following minerals: Calcium; Copper; Iron; Fluorine; Phosphorus; Magnesium; Potassium; Zinc.
- Nutritious and easy to handle in the kitchen, beans are a food that can be

consumed in many different ways. Whether with rice, in farofas, salads, as a broth or in the traditional feijoada, dishes containing beans are tasty and rich in very important substances for the body.

3.2. Conditions climatic

The climate that characterises the region is tropical dry, megathermal, with irregular rainfall in the months of February to June, not exceeding 600 mm per year in an irregular manner, with an average relative humidity of less than 50 %. The climatic conditions of the experimental area correspond to an ecological zone with little vegetation of semi-arid tropical trees and bushes, 35.2 % relative humidity, 15.54 °C minimum temperature and 33.35 °C maximum temperature.

The predominant climate in the region is hot and dry with a rainy period from November to March. Rainfall was nil in the months of the experiment (see Annex 1 for climatic data according to quarters of the year, months of the experiment).

3.3. Design experimental

A Randomised Block experimental design was followed, planting was carried out in independent plots for each treatment and three replications for each of them, making a total of eighteen plots, a sampling system was designed for the variables under study that allowed statistical analysis of the data, processing and discrimination of the most promising variants for local conditions.

The plots were spaced 0.50 m apart. The plots were 2.5 m long and 2.5 m wide, the distance between rows was 0.60 m, the distance between plants was 0.25 m with 40 plants per plot. The way the plots were placed during the experiment to evaluate the agro-productivity response of different doses of organic fertiliser (cattle manure) and the different treatments are shown in table 7.

Table 7. experimental treatment and doses used

Abbreviation	Treatment	Dose
T1	Bean variety	5 kg.m-2
	(Phaseolus vulgaris,	
T2	L.) "waiting for my	10 kg.m-2
T3	brother-in-law "with different	30 kg.m-2
T4	organic fertiliser dosage	Yes implementation.
(Witness)	(livestock manure).	

3.4. Deals

All treatments had the same number of plants, with the same bean variety but with different doses of organic fertiliser. To investigate each experimental plot, the central furrows were chosen.

3.5. Management of experiments

3.5.1. Preparation of soil

The soil was prepared a month before sowing in order to provide a place for sowing that would allow the plants to develop quickly; goat manure was used at the bottom of the furrows, and a week before sowing, the soil was ploughed 20 to 25 cm deep, followed by harrowing to obtain a good soil crunch.

3.6. Sowing and care cultural.

The seeds for bean production were purchased at the Shomocuyo market, they are easy to germinate due to their permeable cover, with a germination power of 95 %. Sowing was carried out on 25 February 2023, at a depth three times the diameter of the seed, in order to allow the seed to germinate easily. weed control Weeding was carried out continuously with an interval of 30 days, ensuring that the soil was free of this pest, in order to avoid damage to the plants during their development. Fertilise The application of goat manure was carried out seven days before sowing directly to the soil by hand, at the bottom of the furrow, with the application of a dose of: five, ten and three $kg.ha^{-1}$ respectively for the treatments.

Irrigation

Irrigation was carried out using the direct irrigation system with the help of a hose connected to an electric pump in alternate periods, by furrow with an application three times a week in the first period and subsequently reduced to twice a week.

3.7. Variables analysed

The field evaluation was carried out in the first two months after sowing, evaluating the growth of the plants, where samples were taken from the experimental plots and the following variables were measured (annex 2):

a) Plant height (cm): The length of the main stem was assessed using a tape measure in cm, considering the measurement from ground level to the apex of the stem. Twenty-four plants per plot were used to obtain the average variable in (cm), which were marked after germination.

b) Stem diameter (mm): The level of the root collar was measured from 24 plants with a gauge graduated in millimetres (mm).

c) Number of leaves per plant (u): Assessed by counting and visually observing the leaves of 24 plants per plot.

d) Number of pods per plant (u): This was assessed by visual observation counting the pods of 24 plants per plot, which were marked after germination to obtain the averages of the variables in units (u).

e) Pod length (cm): Assessed at 40 pods per plot by visual observation count at each harvest, measured with 30 cm ruler).

f) Weight of 100 cements (g): One sample per plot was weighed using a Ciatronic kw 3412 electronic scale with a maximum capacity of (50 kg).

g) Yield per plot net kg.ha $.^{-1}$

IV. RESULTS AND DISCUSSION

The main results obtained in the research are shown in the form of figures and tables, which demonstrate the variations in the aspects studied in each of the treatments.

4.1.Behaviour of growth variables

4.1.1. Plant height analysis (cm)

Plant growth and the resulting dry matter accumulation are directly related to the continuous uptake of mineral nutrients, which only occurs if the plant size increases. When analysing the behaviour of the growth variable plant height, significant differences can be observed in the bean variety under study, as shown in figure 1.

Figure 1. Variable plant height behaviour

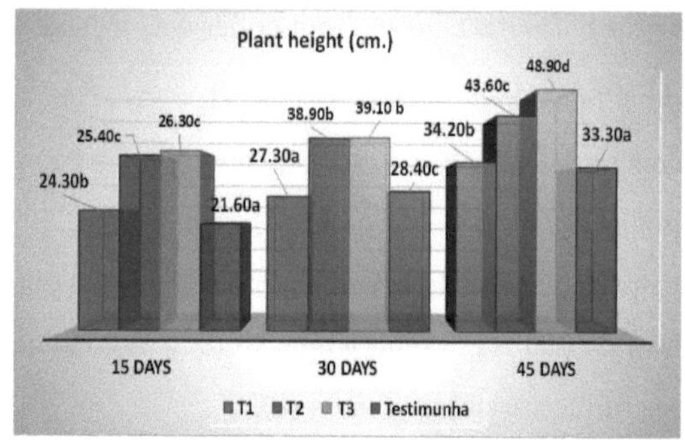

Means with unequal letters (a, b, c, d) differ from each other by p :S 0.05.

When evaluating the effect of the use of different organic fertilisers in a bean variety on the growth variable seed height; it can be seen in figure 1, that the treatment (T3), gives better results with a height of 36.30 cm, at 15 days and at 45 days of 48.90 cm, significant discrepancy with the other treatments and rather differing significantly from this one. treatment (T2) at 30 days, in all cases this result corresponds to treatments when an organic allocation is applied at the rate of 30.0 kg.m^2 , in all cases showing the lowest value in the control treatment (T4). These results are superior to those reported by Sánchez (2010), who evaluated two bean cultivars in the municipality of Ciego da Ávila (Cuba) in a Ferrallitic Red soil, obtaining average values between 23.8 cm and 26.9 cm, respectively. These differences may be due to the environmental conditions in which the research was carried out and the agronomic management to which these crops were subjected, if we take into account what is stated by Palácios and Montenegro (2006), cited by Jiménez (2013), which establishes that plant height is a specific genetic characteristic of the cultivar, which interacts with the environment. Superior results were also reported by Oporta and Rivas (2006) in Nicaragua, when evaluating a bean study to determine ideal densities and sowing time for varieties with indeterminate growth, they found significant differences in the values of main stem length, reaching the highest values of 1.50 and 2 metres. Lower and similar results were reported by different authors, such as Nonato (2007), who evaluated 20 semi-bearing varieties in Brazil under rainfed conditions and found significant differences in these variables with values ranging between 0.36 and 0.48 m.Ndapandula (2013), evaluating two varieties Nacare and Shindimba in three sowing periods, found significant differences in the values of main stem length evaluated at 40 days after sowing (DDS), reaching values of 126 cm in Shimdimba and 198 cm in Nacare. In this research, superior results were achieved by evaluating six bean varieties in a semi-arid soil.

4.1.2. Stem diameter analysis (mm)

The diameter of the stem of the plant is an indicator that, together with the height and number of leaves, form the vigour of the plant, as shown in figure 2. It is observed that the best treatment in terms of the variable stem diameter was in the plants obtained from treatment T4, (where twelve and 30 kg.m^2 of cattle manure) obtained the best results, with a diameter of 8.90, 9.30 and 9.60 mm, at 15, 30 and 45 days respectively, the analysis of variation showed significant differences between all treatments.

Ngouajio et al., (2006) and Oliveira et al., (2010) achieved similar results and report that as the plant gains in growth, the stem increases in diameter. This research shows that this accounts for the potential of each bean variety and a suitable climatic environment for successful plant development.

Figure 2. Variable stem diameter performance

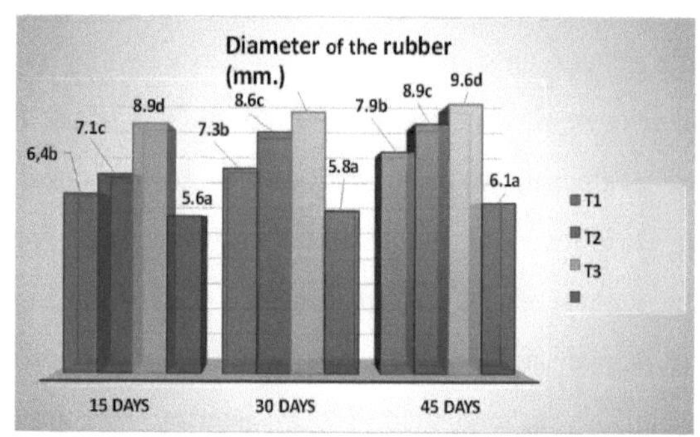

Means with unequal letters (a, b, c, d) differ from each other by p :S 0.05.

Similar results were obtained by Ngouajio et al., (2006) and Oliveira et al., (2010) report that as the plant gains in growth there is a larger stem diameter. This research shows that the above explained is the potential of each bean variety and a suitable climatic environment for successful plant development. Most of the plants in treatments with six bean varieties show that the stem diameter variable was positively influenced by climate and the genetic quality of each variety. The results obtained coincide with the trend reported by Mallor et al., (2010), that a decrease in plant stem diameter occurs when the genetic potential of each variety cannot be expressed in a trial of onion grown in Fuentes de Ebro, Spain.

4.1.3. Behaviour of the variable number of leaves (u)

The increase in leaf area is of great physiological importance for plants, since with the increase of this variable there will be a greater possibility of increasing the active photosynthetic surface, which favours the production of carbohydrates that, together with the water and mineral elements absorbed by the roots, will induce an increase in the synthesis of proteins and other organic compounds, which will translate into an increase in the production of biomass in plants. In the analysis of variance carried out for the variable number of leaves per plant (figure 3), it is observed that there are significant differences between the treatments. The best treatment for the variable number of leaves is observed in the treatment T3 (where dose of organic fertilizer of 30.0 kg.m^2), with values of 9.3; 38.3 and 56.3 leaves, followed by the treatment T2 (7.10; 8.6; and 8.9 leaves) and T1 (6.10; 26.4 and 36.9) in that order respectively, then continues with the control treatment T4 (6.3; 18.3 and 23.6 leaves).

Figure 3. Behaviour of the variable number of leaves per plant.

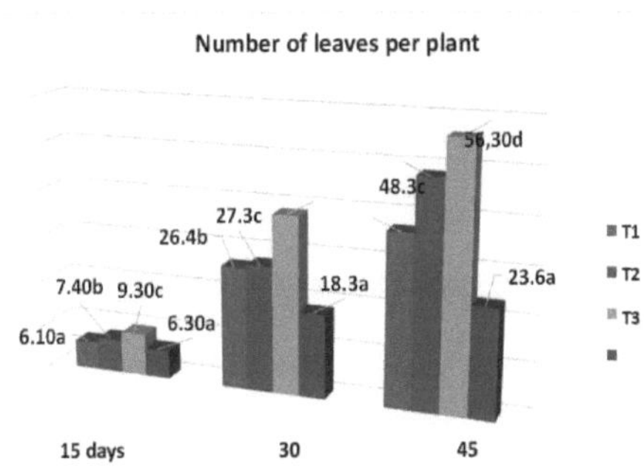

Means with unequal letters (a, b, c, d) differ from each other by p :S 0.05.

These results may be given by the fact of the size of the plant and the degree of deification as this is a character of the plant architecture, and influenced by the characteristics of each variety, plus the environment that allows the development of the plant specifically in the emission of leaves (Manuel, 2014).This research highlighted the beneficial effect caused by the different genetic characteristics of bean varieties, where the plant is less exposed to emissions of solar rays, which, according to Leving (2004) cited by Manuel (2014), radiation in plants is promoting the activation of various enzymes, such as polyphenol oxidases, cabalases, peroxidases and esterases, which support the formation of physiologically active substances, which are used by the plant favouring its growth and development in general. Similar results were reported by Ndapandula (2013), evaluating two bean varieties Nacare and Shindimba in three sowing periods, who found significant differences in the number of leaves per plant, arguing that the difference was due to the variety and behaviour of rainfall, climate and fertilisers incorporated into the soil. Nonato (2007), evaluating 20 varieties of semi-cropped beans of indeterminate growth in Brazil under rainfed conditions, found significant differences in the values of this variable among the different varieties. Filho and Carlos (2006), evaluating nine progenies of macunde bean, found significant differences in the values of main branch length, which ranged from 25.17 to 94.88 centimetres, while the number of branches reached values of 2.60 branches per plant. The number of main lateral branches and their characteristics have a direct influence on plant architecture and grain yield potential, as observed by Mendes et al.

4.2. Behaviour of the components of performance.

4.2.1. Behaviour of the number of pods per plant (u)

The number of pods per plant in the bean crop, together with pod length and 100-seed weight, are the variables that are always related to crop yields, regardless of the conditions evaluated. These results are presented based on the genetic characteristics of the varieties under study, which is an essential component for plant productivity. In figure 4, it is shown that the variable number of pods where or treatment T3 (where organic fertilizer dose of 30.0 $kg.m^2$), obtained the highest results 7.16 pods/plant, differing significantly with all treatments, treatments T1 and T2 and not differing between links, the lowest result is shown in treatment T4 (control). These results do not correspond with those obtained by Sánchez (2010) who reported a minimum value of 9 and a maximum value of 10 pods per plant when evaluating two bean cultivars with I habit. Valdivia (2010) when evaluating five cultivars with habit I reported a minimum value of 9 and a maximum value of 11 pods per plant.

Figure 4. Pod/plant behaviour

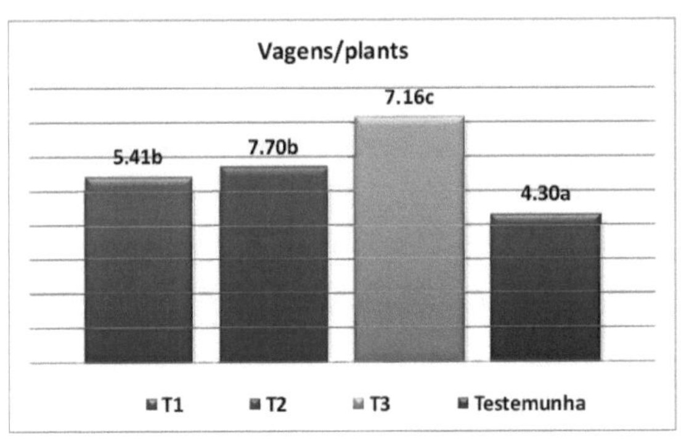

Means with unequal letters (a, b, c, d) differ from each other by p :S 0.05.

Expósito (2010) reported an average value of 13 pods per plant when evaluating a cultivar of black bean habit I in a Ferrallitic Red soil in the municipality of Cego de Avila (Cuba), these values were lower in relation to the values obtained in this research. The number of pods per plant is related to yield, Artola (1990) quoted by Jiménez (2013) states that it depends on the number of flowers that the plant has, in addition, it is influenced by environmental factors (temperature, wind and water), at the time of flowering and nutritional status during the pod and seed deformation phase. The number of pods per plant is discontinuous, as its values can be expressed in whole numbers (Jiménez, 2013).

Many authors have reported similar results regarding the values obtained in this yield component (Aramendiz et al., 2011), evaluating 13 promising cowpea lines in the Sinú valley in Colombia and finding values of 10 and 21.1 pods per plant. Teixeira et al., (2010), in Brazil, evaluating eight cowpea varieties, found values of 10 and 23 pods per plant, these results differ from those obtained in this research with bean varieties.

The purpose of evaluating the performance of different bean varieties in the experimental area is to provide producers with seeds that are resistant to the climatic conditions of Ondjiva, in order to increase yields, take advantage of the plant's living space and improve physical conditions, conditions by increasing the nutritional reserves already in place. existing on the ground. Sowing a bean variety correctly is always one of the most effective means of obtaining the best yields, as well as improving soil fertility. The increasing use of these practices and their application on a large scale are the main elements for increasing bean production in Cunene province. This is associated with low levels of rainfall in the locality, which means that crop yields do not reach their maximum productive potential in the region. In addition to the conditions evaluated, it is necessary to use alternatives that allow high yields to be achieved, while minimising losses due to the incorrect use of varieties in bean cultivation. The length of pods per plant is one of the determinants of yield and is linked to conditions of high intensity of solar radiation due to the increase in leaf area, increasing the photosynthetic capacity of the plant, thus forming nutrients that stimulate the formation of pods. The number of pods and their length on a plant is a genetic characteristic of the variety that changes gradually with environmental conditions (Da Silva Carneiro, 2011). In the analysis of variance carried out for the variable length of pods per bun, it was observed that, in the case of the size of pods per bun, the highest values were reached by the T3 treatment (where doses of organic fertiliser of 30.0 kg.m^2 were applied), significantly different from the other treatments, the lowest value being obtained by the T4 control treatment. Although this pattern is related to productivity, since the higher the number of pods per plant and the higher the number of grains per pod, the higher the productivity Carvalho et al. (2009), carrying out a morphoagronomic characterisation and its genetic divergence in 44 African cowpea studies in the State of Piauí, Brazil, reported values between 10 and 18.9 centimetres in length per pod and from 1.5 to 4.5 pods per stem; in this research, the length of pods per plant in the bean crop, the results were lower than those obtained by this author. Da Silva Carneiro (2011), evaluating six cowpea varieties in Brazil, found pod length values between 12 and 18.3 cm and stated that the pod lengths of all varieties were shorter than the commercial standard. of 20 cm, proposed by Silva and Oliveira (1993), with values higher than those of 20 cm. of this research.

4.2.2. Behaviour of grains per pod (u)

The performance of two kernels/plant is very important when analysing yield components because the more kernels there are in the cobs, the better the performance or yield Da Silva Carneiro (2011).

Figure 5. Grain behaviour by Pods

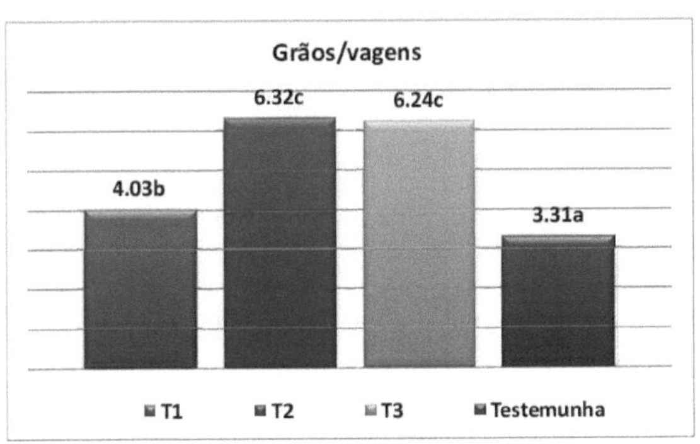

Means with unequal letters (a, b and c) differ from each other by p :S 0.05.

These results do not correspond with those obtained by Sánchez (2010) who reported a minimum value of 9 and a maximum value of 10 pods per plant when evaluating two bean cultivars with I habit. Valdivia (2010) when evaluating five cultivars with habit I reported a minimum value of 9 and a maximum value of 11 pods per plant.

4.2.3. Weight behaviour of 100 seeds (g)

When analysing the weight of 100 seeds in (figure 6) it can be seen that the best performance was achieved by the T4 treatment (control) with a value of 15.3 g, reporting statistically significant differences between all treatments. Treatment T3, with a value of 19.46 g, differs significantly from treatments T1, T2 and T4. Ndanyengwondunge, (2014), explains that the weight of the seed is conditioned by the transfer of nutrients from the plant to the seed, during the its vegetative stage. Exania and Zoraida, (2006) cited by Ndapandula (2013), the weight of 100 seeds is determined by the size of the seed: long and thick. In this research, these authors were taken into account.

Figure 6. Behaviour in the variable weight of 100 seeds (g)

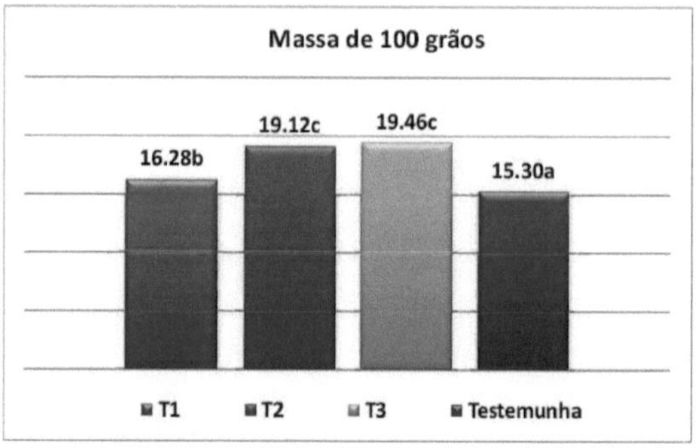

Massa de 100 grãos

16.28b
19.12c
19.46c
15.30a

T1 T2 T3 Testemunha

Means with unequal letters (a, b, c and d) differ from each other by p :S 0.05.

Nonato (2007), evaluated 20 indeterminate growth varieties of size semi-cropped Macunde bean in Brazil under rainfed conditions, reporting values ranging from 12.7 to 25.7 g, indicating high variability of this component among varieties. Da Silva Carneiro (2011), in Brazil, evaluating six varieties of Macunde reported values for the weight of 100 seeds that vary between 18.5 and 22 g. In this research, taking into account the varieties of this crop, the results obtained are higher. .

4.2.4. Yield performance ($t.ha^{-1}$)

When evaluating the yield (figure 7), it is observed that there are significant differences between treatments, with the T3 treatment (where organic fertiliser doses of 30.0 $kg.m^2$ are applied), with a value of 0.96 $t.ha^{-1}$ achieving statistically significant differences between the T1 and T4 treatments, and not with the T2 treatment, the lowest 0.72 $t.ha^{-1}$, differing significantly with all treatments (figure 7).Yield determines how efficiently plants use nutrients. and the genetic potential of the varieties. According to Palácios and Montenegro (2006), yield is a function of several anatomical and morphological characteristics related to the number of pods per branch, number of pods per plant, number of seeds per pod and 100-seed weight.

Figure 7. Performance of equities by share

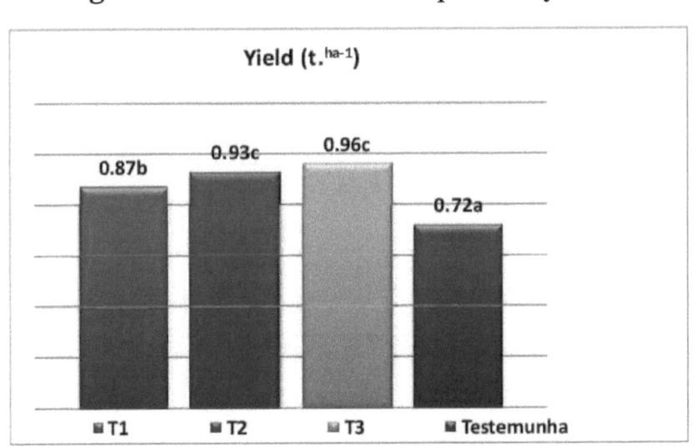

Means with unequal letters (a, b and c) differ from each other by p :S 0.05.

Ndapandula (2013) cited by Ndanyengwondunge (2014) states that yield formation occurs throughout the entire period of growth and development, from the emergence of the plant until the formation of the last organ under the influence of soil and climatic factors, so the evaluation should consider the specific environment in which the experiment is conducted, that the high and low values reflect the real possibilities of the genotype, according to the conditions present. In this experiment, treatment T3 (where organic fertiliser dose of 30.0 kg.m^2) was the variety with the best adaptability to the climatic conditions of the study site. Samaneolu (2013), under similar soil and climatic conditions, evaluating five planting distances with the Shindimba variety of macunde bean, found values of 1.36 and 1.85 t.ha^{-1} , respectively, taking into account the characteristics of the bean varieties studied. under these conditions, these parameters have lower and intermediate behaviours to those obtained in this research.

4.3. Analysis economic

Table 8. Analysis of the economic performance of the bean treatments.

Treatment	Performance (Gross production) (t.ha-1)	Selling price (AKz)	Gross value of production (AKz)	Gross production cost (AKz)	Ganância (AKz)	Cost x Akz. (AKz)
T1	0,87	850.000	739 500,00	34 000,00	705 500,00	0,21
T2	0,93	850.000	790 500,00	32 154,00	758 346,00	0,24
T3	0,96	850.000	816 000,00	36 450,00	779 550,00	0,21
T4 (Witness)	0,72	850.000	612 000,00	21 587,00	590 413,00	0,27
total	3.48	850.000	2 958 000,00	124 121,00	2 833 809,00	0,23

Notice:

✓ Price of one $t.ha^{-1}$ of beans = 850.000,00 Akz.

✓ Value of gross production = Gross Production x price of one $t.ha^{-1}$ of beans.

✓ Profit = Gross Value of Production - Gross Cost of Production.

✓ Cost x Akz = Profit / Gross Cost of Production.

As can be seen in table 7, treatment T3 (where organic fertiliser doses of $30.0kg.m^2$ were applied), obtained a production cost of 36,450.00 Akz and with a sale produces 816,000.00 Akz, guaranteeing a profit of 779,550.00 Akz, with a cost per Akz of gross production of 0.21 Akz, consider this result or better short experiment.

Following this value, the results of treatments T2 (where organic fertiliser doses of $10.0\ kg.m^2$) and T1 (where organic fertiliser doses of $5.0\ t.ha^{-1}$) are shown, the production costs in each case were: 32 154.00 Akz and 34 000.00 Akz, the lowest costs were found in treatment T4 where the control was sown, although it shows the highest costs. results of their cost of production (0.27 Akz). In all cases there was a gain and, in general, the Cost x Akz was fast at 0.23 Akz. Ndapandula (2013), evaluating two varieties Nacare and Shindimba of macunde bean in three sowing periods, found a better cost x Akz ratio in December sowing due to the higher yield obtained, in this research the best results were obtained by treatment T3 (where dose of organic manure $30.0\ kg.m^2$) and where the best economic index comes from treatment T3 (where dose of organic manure 10.0 kg.m).[2]

V.CONCLUSIONS

1. In the agro-productive performance of the different organic fertiliser applications, bean varieties and treatment T3 (where organic fertiliser doses of 30.0 $kg.m^2$ are applied) obtained the best results in the variables plant height, stem diameter and number of leaves per plant.

2. The best bean yields were achieved in the T3 treatment.

(where organic fertiliser doses of 30.0 $kg.m^2$ were applied).

3. Economically, the best treatment was T3 (where organic fertiliser doses of 30.0 $kg.m^2$ were applied), yielding a profit of 779,550.00 Akz, although all treatments made a profit.

VI. RECOMMENDATIONS

1. That farmers in the province of Cunene, in particular those in the locality of Caxilla III, apply the techniques available in this work to achieve better yields in the cultivation of beans, using organic fertiliser applications at a dose of 30.0 $kg.m^2$, as this is the treatment that best develops in the local climatic conditions.
2. Carry out the same study in other municipalities of the province with different soil and climate characteristics to test the adaptability and behaviour of different bean varieties.

VII. BIBLIOGRAPHY

Almaguel, A. (2002). Taxonomy and Phytosanitary Diagnosis of mites of agricultural importance. Havana City, CUBA.

Altieri, MA (1999). Agroecology. Scientific basis for agriculture. sustainable. Editorial Comunidad Nordana. Archive. Peru.

Arteaga, L. and Pumalpa, N. (1992). Situación actual del manejo de la cosecha de frutas y hortalizas en el Departamento de Nariño. In: Seminario sobre manejo poscosecha y comercialización de frutas y hortalizas. Pasto, Colombia. p.1-8.

Araya, CM (2010). Guía para la identificación y manejo integrado de enfermedades del frijol en Centroamérica/ IICA/ Proyecto Red SICTA, COSUDE. Managua. Nicaragua.

Arias, JH; Restrepo, T. and Jaramillo, Carmona, M. (2009). Manual: Good Agricultural Practices in Bean Production.

Artola, EA, (1990). Effect of furrow spacing and disease control in common bean (Phaseolus vulgaris L.), genotype Revolution 81 in spring cycle 1988. Thesis of the National Agrarian University (UNA). Managua, Nicaragua. 1 37p.

Ayala, JL and Mar, R. (2014). Phytosanitary manual for urban and pre-urban agriculture. FAO Bulletin. South-South Cooperation. Caracas, Venezuela.p.26.

Baudoin, JP; T. Vander Boght; K. Mwangombe (2021). Crop production in Africa. Edited by Romain H. Raenoekers. DGIC. Brucellas. p. 31 7-334.

Carpeta, U. (2007). Manual de leguminosas nicaragüenses. Volume I. Estelí, Nicaragua. 191p.

Box, JM (2011). Grain legumes. Ed. Salvat. Barcelona. 550 p.

Cabrera, C. (2000). Frijol. You can live in Ecópolis. Fundación Antonio Núñez Jiménez de la Naturaleza y el Hombre, Vol.,5, núm.20,8-11.

Camellón, M. (2012). Incidence (Polyphagotarsonemuslatus Banks) in bean cultivars (Phaseolus vulgaris. L) and regulation of its problems. Thesis in option for the degree of Agricultural Engineer. University of Ciego de Avila. Cuba.

Christmas letters. EL/Cities/Angola. Geographical coordinates and area Hourly Cities from Angola. 2012. Available<http://www.google.com/Cartasnatal.es/ciudades/Angola- p 795. Accessed 3/3/2023.

Carvalho, A.; Bezerra, A.; Alves, FJ; Fernández, T.; Rodríguez, F.; Freire F. and Queiroz, Ribeiro, V. (2009). Characteristics of cowpea yield rates at different stocking densities. Agropec. Bras, 2009, Brasilia, v.44, n.10, p.1239-1245.

Camagro, (2005). Cámara Agropecuaria y Agroindustrial de el Salvador. Manual para el manejo postcosecha de hortalizas. p.19. [accessed on 13 May 2023]. Available on the World Wide Web: <https://www.bmi. gob.sv/pls/portal/url/ITEM/1FFDE4C4A5A0D9E8E040558CE3C96 EF2

International Center for Tropical Agriculture, (CIAT). (2014). Morphology of the common bean plant (Phaseolus vulgaris L.). Study guide. CIAT, Cali (Colombia), 49.

International Center for Tropical Agriculture, (CIAT). (2010). Year descriptions of pests attacking beans. Study guide. CIAT. Cali, Colombia. 41 pp.

Cossanga, E. (2000). Effects of some sustainable alternatives on crop development and growth. Diploma thesis. Granma University, Bayamo.

Da Silva, Carneiro, A. (2011). Agronomic characteristics and quality of cowpea seeds in Vitória da Conquista, BAHIA. Thesis presented at the State University of Southwest Bahia as part of the requirements of the Postgraduate Programme in Agronomy, area of concentration in Plant Science, for the Master's degree. Vitória da Conquista Bahia-Brazil. 2011.

Escoto, D. (2014). The cultivation of beans. Technical manual for use by private companies, individual consultants and producers. Secretariat of Agriculture and Livestock. Directorate of Agricultural Science and Technology. Tegucigalpa. Honduras.

Cunene Meteorological Station (EMC) (2017): Mapping information on temperatures, relative humidity and precipitation. Cunene Meteorological Station. National Institute of Geophysical Hydrometeorology, Ondjiva.

Exania de Rosario, JJ and Zoraida de Socorro, L. (2006). Preliminary evaluation of 36 genotypes of common bean (Phaseolus vulgaris L.) in a later season in Macico, Somoto. Diploma thesis. Managua. Nicaragua, 2006.

Expósito, Y. (2010). Evaluation of cultivars of red bean (Phaseolus vulgaris L.), in a Ferrallitic Red soil in Ciego de Avila. Work as optional Agronomist Engineer. Cuba.

Fandiño, Y. O. (2009). Study and regionalization of common bean varieties INIVIT Punti blanco (Phaseolus vulgaris L.), in the Central Eastern Zone of Cuba. Optional Master's thesis in Agricultural Sciences. University of Ciego de Avila. Faculty of Agricultural Sciences. Cuba.

FAO. (2011). Agricultural production, primary crops. Available at: <http://faotatz.fao.org/home/index_es.html.locale-es#.

FAO. (2013). Criteria, tools and means for agriculture and food security. sustainable rural development. Main document nº1. Granada.

FAO. (2015). Main imports of dry beans in 2015. FAO Statistics Division.

FAOSTAT, (2010). Maximum importance of dry beans in 2010.

FAO Statistics Division.

Filho Aires, M. and Carlos, H. (2006). Genetic analysis of characters related to plant architecture in cowpea (Vigna unguiculata L. Walp). Teresina Estado de Piauí - Brazil, p 32.

Franco, F.; Pedroso, R.; Noá, A.; Castañeda, I.; Ríos, C.; Arredondo, I. and Chacón. R. (2014). Official list of plants. Complementary material for Botany. Centro de Estudios Jardín Botánico de Villa Clara. UCLV. SantaClara. Villaclara.

Flores, Elisa (2012). Response of tomato (Solanumlycopersicum. L) crop with the use of different sources of organic matter, in semi-desert conditions of Gtmo. Work to obtain a Diploma in Option to the Degree of Agronomist Engineer. Cuba

García, F.; Costa, J. and Laborda, R. (2009). Agricultural pests. Exopterygote insect mites. Polytechnic University of Valencia Spain.

Hernández, R. (2012). Biological regulation of Empoascakraemeri Ross and Moore, Harrisemum bean cultivar (Phaseolus vulgaris L.), at CPA "26 de julio". Thesis presented as an option to Agronomist Engineer. Cuba.

Hernández P. and Chailloux S. (2001). Influence of different sources of organic benefits on growth and nutrient use in Vignaluteola. Programme. Res. XIII Scientific. Scientific. INCA, p. 67, Havana.

IMPULSEMILLAS, (2007). Hybrid cabbages. En: Product catalogue. Seeds and vegetables line. [online]. [cited 20 May 2006]. Bogotá.

Available at on World Wide Web:

<http://www.impulsemillas.com/cat_product.asp?id_p=37&id_c=16 Infoagro (2008). Vegetable cultivation. Available at: http://wwwinfoagro.com/hortaliza/col.htm. Accessed on: 30 May 2023.

Jaramillo, J and Leyva, E. (2002). The cultivation of cruciferous vegetables (Cabbage - Broccoli - Cauliflower). In: Vegetable Productivity-Marketing Workshop. Tibaitatá, CORPOICA. pp.14-32. [online]. [21 June 2021]. Bogotá. Available at World Wide Web:

<http://www.corpoica.org.co/Archivos/Foros/MEMORIAS.pdf.

Jiménez, S. F. (2013). Generation of pest signalling methodologies.p.20-28. In: LL Vázquez; Ingrid Paz (eds.), Memorias del Curso Taller para Agricultores y Extensionistas "Manejo integrado de plagas en la producción agrícola sustentable". INISAV, Havana, Cuba.

Lee, B. (2001). Evolution of some biotic, abiotic and lean ecosystem factors of bean (Phaseolus vulgaris L.) associated crops in tkkkt (Zea mays). Optional thesis for the Master's degree in Agroecology and Sustainable Agriculture.

Agrarian University of Havana, Havana, Cuba: 67p.
Leving, V. (2004). Dynamic accumulation of generic substances and variations in protein quality in wheat plant stocks from semi-irradiated plants. Agricultural Biology. University and Science Journal. 2004, vol. 20, no. 39, p. 52-57.
López, R. (2014). Biological peculiarities of (Andrectorruficornis Oliver) in a bean crop (Phaseolus vulgaris L.) in a field in the municipality of Victoria, Venezuela. Thesis presented as an option for Agronomist Engineer. UNICA, Cuba.
Manuel, D. (2014). Effect of sowing distance on the agricultural yield of cucumber (Cucumissativus LV Ashley). Final course work submitted for the degree of Agronomist Engineer. Instituto Superior Politecnica do Cunene-Angola, 2014.
Martínez, E.; Barrios, G.; Rovesti, L.; Santos, R. (2007). Integrated pest management. Manual Práctico, Centro Nacional de Sanidad Vegetal, Instituto de Investigaciones Fitosanitarias, Havana, Cuba.
Mendes, RM de S; Távora, FJA F; Pinho, JLN and Pitombeira, JB (2005). Changes in the source-sink relationship in cowpea subjected to different plant densities. Revista de Ciencias Agrícolas, v.36, p 82-90.
MINADOR (2015). Review of results of the 2014/2015 agricultural campaign of the Ministry of Agriculture. Angola.
Ministry of Agriculture (2016). Agricultural sector outlook for 2016. Available at: http://www.embajadadeangola.com/embajadadeangola- Angola-economia.hTML.Accessed 26 September 2021.
Ministry of Agriculture (2019). Characterisation of the Province of Cunene. Report of the Provincial Forestry Department of Cunene. Luanda: Ministry of Agriculture. Angola.
Mora, A. (2013). Origin and importance of the common bean crop (Phaseolus vulgaris L.). Revista Facultad de Agronomía de Maracay, vol. 23, 225- 234.
Namwenyo, J. (2017). Inventory of harmful organisms affecting Bean

(Phaseolus vulgaris L.) production in Ombandja municipality. Management proposal. Final course work submitted for the degree of Agronomist Engineer. Polytechnic College of Cunene, Ondjiva.

Ndanyengwondunge, CJ (2014). Performance of three improved varieties of Macunde bean (Vignaunguiculiculatawalp Lin) under rainfed conditions in Cunene province. 59 ff. Dissertation (Degree in Agricultural Engineering) - Escola Superior Politécnica do Cunene, Ondjiva.

Ndapandula, Lourenço, M. (2013). Performance of two improved cultivars of Macunde bean (Vignaunguiculata Lin) at three sowing periods. Final course work submitted for the degree of Agronomist Engineer. Mandume ya Ndemufayo University, Ondjiva Cunene-Angola Polytechnic.

Ngouajio, M.; Wang, G. Y. and Haus Beck, M. K. (2006). Changes in yield and economic value of pickled cucumber in response to planting density. Crop Science. 2006, vol. 1, no. 46, pp. 1570-1575.

Nonato, BR (2007). Evaluation of genotypes of semipropped cowpea in cultivation.

rainfed and irrigated. TERESINA. PI. May 2007, 37p.

Oliveira, A.; Silva, J.; Oliveira, AN; Silva, D.; Santos, R. and Silva, N. (2010). Gherkin production as a function of inter-row and inter-plant spacing. Brazilian Horticulture. 2010, No. 28, p. 344-347.

Oporta, Pichardo, E. and Rivas, Cáceres, AM (2006). Effect of population density and dry season on yield and seed quality of a population of red cowpea (Vigna unguiculata L. Walp) in the El Plantel farm. Thesis presented for the option of agronomist engineer at the National Agrarian University. Managua, Nicaragua October 2006, 31 p.

Pacheco, D. (2011). Bean demand for improved technology. (Talk presented at an International Conference, Cali, Colombia). Mimeo 1 4p.

Palácios, A. and Montenegro, D. (2006). Effect of three drying densities and three drying seasons on growth and yield of red cowpea (Vigna unguiculata L.

wal). Thesis work. Ciudad Darío, Matagalpa. National Agrarian University (UNA). Managua, Nicaragua. 46p.

Small TC (2018). Evaluation of six bean varieties (Phaseolus vulgaris. l) under Ondjiva climatic conditions. End of course work submitted for the award of Bachelor of Engineering degree. Cunene Polytechnic Institute. Department of Engineering. University Mandume Ndmufayo.

Ramos, J. (2014). Pests and diseases of economic importance affecting the bean crop in the Chincha valley. Regional Government of ICA. Dirección Regional Agropecuaria del ICA. Chincha Agrarian Agency. Costa Rica.

Rodríguez, Y. (2006). Evaluation of 15 cultivars of red bean (Phaseolus vulgaris, L) under soil and climatic conditions in the municipality of Majibacoa (option as agronomist engineer). Las Tunas University Centre. Cuba.

Samaneolu, P. (2013). Seeding density in rainfed cultivation of macunde bean (Vignaunguiculata L. Walp). Final course work submitted for the degree of Agricultural Engineer. University of Mandumeya Ndemufayo, Escola Superior Politécnica de Ondjiva Cunene- Angola, 2013.

Sánchez, R. (2010). Agro-productive behaviour of 19 bean varieties (Phaseolus vulgaris L.) in a red Ferrallitic soil in Ciego de Avila. Work as an option to the degree of Agronomist Engineer. General Máximo Gómez Báez" University of Ciego de Ávila. Cuba.

Sandoval, P. (2010). Plant health in three bean varieties (Phaseolus vulgaris L.) under favourable and unfavourable soil moisture conditions. Revista Chapingo, Serie Zonas Árida.

Sandoval, P. (2014). Response of bean varieties to treatments with organic and chemical fertilizers in the control of the main bean diseases in a Lagunera region. Rev. Mexicana de Fitopatología 12:63-76.

Silva, PSL and Oliveira, CN (1993). Yields of green and mature beans from cultivars of cowpea (Vigna unguiculata L. Walp). Brazilian Horticulture, v. 11, p. 133-135.

Help, Q.; Martin, F.; S, David (2008). Grains. Printed at the Figuras de Oficinas del Departamento de Publicaciones y Materiales Educativos del Instituto Politécnico Nacional. Trasgueras 27 Centro Histórico, México, DF.318p.
Mota Yumeri (2012). Cucumber (Cucumis sativa) crop response with the use of organic alternatives under intensive greenhouse conditions. Optional Diploma thesis for the degree of Agronomist Engineer. Mountain Agroforestry Faculty. Subsede Costa Rica. Ministry of Higher Education. Guantánamo.
Tapia, B. (2014). Common bean (Phaseolus vulgaris L.). Ministerio de Desarrollo Agropecuario y Reforma Agraria. Nicaragua: 42-55. pag.
Teixeira, Rhode Island; Carneiro, Da Silva, G.; Ribeiro, De Oliveira, JP; Guerra, Da Silva, A. and Pelá, A. (2010). Agronomic performance and seed quality of cowpea cultivars in the cerrado region. Revista. Ciencias agronómicas, vol.41. norte. .2 Fortaleza.
Tortosa, G., Alburquerque, J., Ait-Baddi, G., Cegarra, J. (2012). The production of commercial organic amendments and fertilisers by composting biphasic olive mill residues ("alperujo") Journal of Cleaner Production, 26, 48-55 DOI: 10.1016/j.jclepro.2015.12.008.
Travuco, M.; Zarza, H.; Morel, L. (2016). Evolution of chemicals and plant extracts for white mite (Poliphagtarson emuslatus Bank.) management in pepper (Capsicuman nuum L.). RevistaTecnologíaAgraria.2; 1 (1):1 5-1 9. Nicaragua.
Valdivia, G. (2010). Agro-productive behaviour of 19 varieties of red bean (Phaseolus vulgaris. L) in a red Ferrallitic soil in a province of Ciego de Avila. Work in option for the degree of Agronomist Engineer. General Máximo Gómez Báez" University of Ciego de Ávila. Cuba.
Vázquez, L. (2008). Integrated Pest Management. Questions and answers for extensionistas y agricultores de La Habana. CIDISAV, 560 p.
Vázquez, LL; Murguido, C. and Peña, E. (2011). Biological control for the conservation of natural enemies in introduced pest management programmes. In: Abstracts IV Scientific Seminar on Plant Health, Workshop on Emerging Pests.

Matanzas, Cuba.p. 257.

Yalta, J. (2001). "Effect of Mulch with chicken incorporation on the yield of cabbage (brassica oleracea, Var. Capitata alba L.). Thesis for the degree of Agronomist Engineer.

Borrero, Y. (2005). Effect of different doses of organic matter on the cultivation of tomato (Lycopersicon esculentum M.), variety HA - 3057, under Protected Growing House conditions. 52h. Diploma work (optionally for the degree of Agronomist Engineer). Granma University.

VIII. ANNEXES

Annex 1. Climate data

MINISTÉRIO DAS TELECOMUNICAÇÕES, TECNOLOGIA DE INFORMAÇÃO E COMUNICAÇÃO SOCIAL
INSTITUTO NACIONAL DE METEOROLOGIA E GEOFISICA
DEPARTAMENTO PROVINCIAL DE METEOROLOGIA DO CUNENE

MAPA DE TEMPERATURAS MÁXIMA E MINIMA, HUMIDADE RELATIVA E PRECIPITAÇÃO REFERENTE AO I TRIMESTRE DE 2023 – DADOS CLIMATICOS

Dias	Tª Máxima °c			Tª Mínima °c			Humidade Relativa %			Precipitação mm		
	Jan	Fev	Marc	Jan	Fev	Marc	Jan	Fev	Marc	Jan	Fev	Marc
01	30,3	29,2	33,1	19,3	20,0	19,7	70	61	51			
02	30,4	30,9	29,8	21,0	21,5	19,1	61	51	66			
03	23,7	30,1	32,8	19,8	21,3	19,8	67	67	44			
04	20,1	32,1	33,0	18,7	20,8	18,0	98	48	20			
05	29,8	31,4	34,0	19,0	22,2	18,4	40	30	22			
06	31,1	30,9	35,9	19,6	20,8	18,8	29	59	21			
07	33,7	26,7	35,9	20,8	20,9	21,4	31	58	35			
08	32,3	28,6	31,0	22,9	21,7	21,5	27	65	50			
09	31,3	31,7	34,7	21,2	19,4	22,6	30	43	30			
10	33,2	33,1	34,4	23,1	18,8	22,4	44	46	40			
11	28,9	33,1	32,0	21,2	20,0	20,8	75	31	78			
12	31,6	33,8	32,8	21,6	20,3	22,6	67	31	68			
13	30,7	35,1	34,8	19,9	21,7	21,0	62	35	40			
14	31,1	35,6	35,0	18,0	23,4	21,2	68	31	39			
15	31,9	35,2	34,3	18,0	21,6	21,1	50	22	33			
16	29,6	33,3	33,4	20,7	18,2	20,2	71	27	47			
17	26,4	32,9	31,7	20,0	16,9	17,7	67	31	40			
18	25,7	31,9	33,5	19,0	13,7	20,0	85	23	50			
19	25,1	33,0	33,3	18,1	15,0	19,4	96	22	46			
20	22,3	33,1	33,9	20,6	15,4	19,3	90	24	49			
21	27,2	34,2	32,6	18,6	14,8	20,2	78	22	55			
22	24,6	35,4	32,4	17,2	17,4	21,3	71	26	47			
23	27,4	34,0	29,8	19,4	21,3	18,8	79	39	72			
24	28,4	32,9	32,0	18,5	20,8	20,5	79	48	53			
25	30,2	34,5	30,8	20,6	20,5	18,5	68	49	46			
26	31,0	34,3	25,6	19,0	20,2	21,2	71	46	84			
27	31,4	32,5	30,6	20,2	21,6	18,3	70	64	55			
28	30,1	27,9	28,4	21,1	19,7	18,6	68	-	77			
29	30,3		31,0	20,1		20,3	59		66			
30	27,8		31,8	17,9		20,6	77		63			
31	27,4			19,6		18,9	76		62			
Total	901,6	906,3	1006,6	548,8	548,8	622,2		1099,0	1565,0			
Média	28.6	30,2	32,5	19,7	19,6	20,1	65	78	51			

MAPA DE TEMPERATURAS MÁXIMA E MINIMA, HUMIDADE RELATIVA E PRECIPITAÇÃO REFERENTE AO II TRIMESTRE DE 2023 – DADOS CLIMATICOS

Dias	Tª Máxima °c			Tª Mínima °c			Humidade Relativa %			Precipitação mm		
	Abril	Maio	Junho	Abril	Maio	Junho	Abril	Maio	Junho	Abril	Maio	Junho
01	31,1	33,0	31,5	20,2	16,7	13,8	61	30	25			
02	32,0	32,5	32,1	19,4	15,1	15,5	59	27	29			
03	32,2	33,0	31,3	20,6	14,4	13,0	50	30	27			
04	31,8	32,5	32,7	20,6	14,4	12,4	55	31	22			
05	31,2	32,3	31,8	18,8	16,5	11,4	43	30	21			
06	32,6	32,6	31,8	17,8	16,5	11,4	54	25	13			
07	33,8	33,1	31,6	18,0	15,2	13,4	45	31	17			
08	33,4	32,8	30,9	19,6	17,8	12,6	39	35	21			
09	32,6	32,8	30,4	20,3	16,3	13,4	42	40	19			
10	33,6	33,5	31,2	21,3	16,3	13,4	46	24	20			
11	33,9	32,8	30,7	19,5	18,3	11,4	34	26	20			
12	33,4	31,7	27,2	19,5	15,1	11,8	40	30	22			
13	34,4	31,8	29,2	18,0	15,9	7,8	39	37	17			
14	33,7	31,7	29,4	19,8	14,2	9,2	35	28	24			
15	33,9	31,5	29,4	18,5	15,0	9,3	37	21	21			
16	33,4	30,2	30,1	16,6	19,9	7,7	33	34	15			
17	32,6	30,3	30,1	17,7	13,4	7,8	37	30	15			
18	32,7	30,5	30,0	16,7	11,2	9,2	45	24	18			
19	33,4	31,0	29,9	18,7	11,3	7,3	46	25	22			
20	31,6	32,3	30,0	20,2	12,3	8,1	45	18	22			
21	20,9	32,4	29,7	17,9	13,3	10,8	75	20	22			
22	29,2	33,1	29,6	18,0	15,6	10,8	72	25	20			
23	32,5	32,4	29,3	19,4	14,5	9,0	45	20	21			
24	32,7	32,4	28,3	20,6	16,4	11,2	46	28	22			
25	32,6	31,1	28,5	20,8	15,2	11,0	56	23	27			
26	33,3	31,7	28,5	19,1	13,4	9,5	48	21	34			
27	33,6		23,0	18,2	15,1	9,4	39	23	52			
28	33,5	31,7	22,4	18,2		3,3	28		32			
29	33,7	31,9	27,1	16,4	15,9	3,6	35	20	21			
30	33,4	29,8	28,2	16,0	12,6	6,8	33	23	27			
31		30,7			11,5			22				
Total	973,6	959,1	885,9	566,4	449,3	305,3	1362	801,0	661,0			
Média	32,4	31,9	29,5	18,9	14,9	10,1	45	27	22			

Annex 2. Information gathering at the Cristiano Shafodino mining company

Variables data collection.

Printed by Books on Demand GmbH, Norderstedt / Germany